International Conference on Material Engineering and Advanced Manufacturing Technology (MEAMT 2017)

IOP Conference Series: Materials Science and Engineering
Volume 282

Busan, South Korea
25 - 27 August 2017

ISBN: 978-1-5108-5499-4
ISSN: 1757-8981

Printed from e-media with permission by:

Curran Associates, Inc.
57 Morehouse Lane
Red Hook, NY 12571

Some format issues inherent in the e-media version may also appear in this print version.

Printed by Curran Associates, Inc. (2018)

For permission requests, please contact the Institute of Physics at the address below.

Institute of Physics
Dirac House, Temple Back
Bristol BS1 6BE UK

Phone: 44 1 17 929 7481
Fax: 44 1 17 920 0979

techtracking@iop.org

Additional copies of this publication are available from:

Curran Associates, Inc.
57 Morehouse Lane
Red Hook, NY 12571 USA
Phone: 845-758-0400
Fax: 845-758-2633
Email: curran@proceedings.com
Web: www.proceedings.com

Table of contents

Volume 282

2017 International Conference on Material Engineering and Advanced Manufacturing Technology

25–27 August 2017, Busan, Korea

Accepted papers received: 4 December 2017
Published online: 14 December 2017

Preface

Papers

IOP Conf. Series: Materials Science and Engineering **282** (2017) 011001 doi:10.1088/1757-899X/282/1/011001

PREFACE

The 1st International Conference on Material Engineering and Advanced Manufacturing Technology (MEAMT 2017) was held in Busan, Korea during August 25-27, 2017.

The conference programs covered keynote, invitation, oral and poster presentations from scientists working in similar areas to establish platforms for collaborative research projects in this field, and to find potential opportunities for international cooperation. These conferences covered recent trends and progresses made in the field of material engineering, advanced manufacturing technology and MEMS, nanotechnology and precision engineering. The major goal and feature of the conference is to bring academic scientists, engineers, industry researchers together to exchange and share their experiences and research results, and discuss the practical challenges encountered and the solutions adopted. Distinguished professors and experts in the area delivered the latest knowledge in their respective expertise fields. International scientific activities, like conference, symposium and workshop are big scientific platforms for knowledge exchange among the scientists from all over the world.

The theme of the conference was the key issues associated with science and technology in this rapidly evolving field of research and to promote contact between basic researcher and technological needs for real advanced materials, nanotechnology, precision engineering and industrial applications.

These event was intended valuable vehicles for the dissemination of research results on latest advances made in the area of advanced materials and technological processes, which gave an excellent opportunity for the scientists and researchers around the world to have a common platform to exchange their findings and to discuss the developments and make a base for collaborations in the national and international level.

The volume consists of the selected papers presented at MEAMT 2017. We hope the volume will offer the readers a good opportunity the access to the future directions of researches in this area of materials.

We want to express our gratitude to all the members of conference committee, and reviewers who spared their valuable time, for their advice which have certainly helped to improve the quality, accuracy, relevance and the sincere efforts to maintain the quality of each paper selected for conference program and volume for publication. Also, we would like to thank every author and participant who has contributed to the success of the conference, and we truly hope to see you again in the next year of MEAMT.

Prof. Osman Adiguzel, Firat University, Department of Physics, Turkey
On behalf of the Conference Committee

IOP Conf. Series: Materials Science and Engineering **282** (2017) 011002 doi:10.1088/1757-899X/282/1/011002

Peer review statement

All papers published in this volume of *IOP Conference Series: Materials Science and Engineering* have been peer reviewed through processes administered by the proceedings Editors. Reviews were conducted by expert referees to the professional and scientific standards expected of a proceedings journal published by IOP Publishing.

Digital image processing techniques for the analysis of fuel sprays global pattern

Rami Zakaria[1], Peter Bryanston-Cross[2] and Brenda Timmerman[2]

[1]Department of Mechanical System Engineering, Hansung University, Seoul, S. Korea
[2]School of Engineering, Warwick University, Coventry, UK

*E-mail: dr.rami.zakaria@gmail.com

Abstract. We studied the fuel atomization process of two fuel injectors to be fitted in a new small rotary engine design. The aim was to improve the efficiency of the engine by optimizing the fuel injection system. Fuel sprays were visualised by an optical diagnostic system. Images of fuel sprays were produced under various testing conditions, by changing the line pressure, nozzle size, injection frequency, etc. The atomisers were a high-frequency microfluidic dispensing system and a standard low flow-rate fuel injector. A series of image processing procedures were developed in order to acquire information from the laser-scattering images. This paper presents the macroscopic characterisation of Jet fuel (JP8) sprays. We observed the droplet density distribution, tip velocity, and spray-cone angle against line-pressure and nozzle-size. The analysis was performed for low line-pressure (up to 10 bar) and short injection period (1-2 ms). Local velocity components were measured by applying particle image velocimetry (PIV) on double-exposure images. The discharge velocity was lower in the micro dispensing nozzle sprays and the tip penetration slowed down at higher rates compared to the gasoline injector. The PIV test confirmed that the gasoline injector produced sprays with higher velocity elements at the centre and the tip regions.

1. Introduction

The function of a fuel injector is to deliver individual pulses of fuel into the combustion chamber of an internal combustion engine. Liquid fuels are turned into sprays by increasing the relative velocity between the fluidic phase and the surrounding ambient. This technique is referred to as 'fluid atomisation'. In pressure-based injectors, the fuel is pressurised into a narrow orifice. The average droplet size of the resultant spray is directly related to the design and quality of that atomiser and on the applied pressure; the more effective a fuel atomiser, the smaller droplets it produces. By reducing the mean droplet diameter, the overall surface area of the fuel increases, which improves the evaporation rate and the burning range. This leads to higher rates of released power during the combustion process and lower exhausted pollutant emissions. Beside the droplet size, other global parameters are of importance to engine manufacturers, such as the spray development velocity and the spray pattern. Understanding the behaviour of fluid sprays, therefore, is vital for designing fuel injectors [1-4].

Spray formation process is a complex problem with many parameters involved, such as nozzle geometry and design [5-10], pressure [4,11,12], cavitation [13,14], evaporation [15], viscosity [16-18], surface tension [18], and temperature [19]. Optical techniques provide invaluable information about spray characteristics. Among all optical methods, imaging methods are dominant in the area of experimental characterization thanks to the advances in digital imaging. Modern digital cameras produce high resolution images that can be enhanced and processed in order to characterise the fluid sprays. The focus of most researchers is on diesel and gasoline direct injection techniques under very high-pressure levels (higher than 100 bar). This paper presents visualisation and processing techniques for the analysis of a kerosene-based (aviation) fuel. It discusses the spray global characteristics and development process against line-pressure and time after-start-of-injection (ASOI).

Small rotary engines are useful in various high power-to-weight applications today, such as racing vehicles, robots, and unmanned aerial vehicles (UAVs). In this type of engine, understanding the spray development process can help in optimizing the combustion process. This becomes especially necessary when the time available for atomisation, vaporization and mixing with air is relatively short. We studied the manifold (indirect) injection strategy in a high-speed (15,000 -18,000 rpm) rotary (Wankel) engine. Due to the small manifold size and the high injection rate, timing the injection pulse (against rotor angle and ignition spark) was crucial. The engine was required to operate on (JP8) jet-fuel and line low-pressure (less than 10 bar). Very little work has been reported on both low-pressure injection and kerosene-like fuels in literature.

We developed a visualisation system and an image processing procedure for spray characterization. Our experiments on heavy jet fuel injectors yielded to around 2.4 Tb of data in form of digital images. Therefore, the image processing techniques needed to be planned carefully to present the outcomes of our analysis in an efficient manner.

2. Experimental conditions

Aviation (JP8) fuel sprays were produced by two injectors using a maximum line-pressure of 10 bar into an unpressurised ambient. The first injector was a standard 12 V electric gasoline injector (single hole, Do=584 μm, f_{max}=0.4 kHz, P_{max}=5 bar). The second injector was a high-frequency micro dispensing nozzle (single hole, Do=255 μm, f_{max}=2.2 kHz, P_{max}=10 bar), which was originally developed for medical purposes and never used for fuel injection before. JP8 is commonly used in turbine-engine aircrafts. We have found that a JP8 fuel drop 35 μm in diameter needs approximately 1.2 ms to start evaporating at 500 °C in atmospheric pressure. The temperature of the manifold (in our case and in most of the similar rotary engines) does not exceed 50 °C when the engine is hot, and it is much lower in the cold-start case. Therefore, it is safe to assume that all the evaporation happens inside the combustion place rather than in the manifold. In rotary engines, the fuel mixture is "sucked into" the combustion place by the pressure difference created by the rotor motion. The mixture is then pressurized and ignited, typically by multiple spark plugs. It is important that the fuel is well mixed with air before the suction stage. This study investigates the fuel spray development process inside the injection manifold, just before introducing the fuel mixture into the combustion place. A sufficient penetration of the liquid fuel within the injection manifold can improve the fuel-air mixing. But if a spray 'grows' faster than the fuel intake rate, this may increase the wall wetting, and hence, the unburned hydrocarbons. Therefore, understanding the spray development process is useful for engine design, and also for the development of computational models for engine performance optimization [4]. The engine was a single rotor Wankel engine, which weighed less than 5 kg and produced around 8.5 hp (6.3 kW) of power at 9200 rpm. The small physical dimensions of the engine introduced challenges in terms of producing a fast evaporation rate and atomising using low pressure levels.

Spray images were acquired in a dark isolated cell where all devices were remotely controlled. Strict health and safety regulations were applied due to the explosive nature of the fuel and the risk of direct exposure to class-4 and class-3 laser beams.

3. Visualisation system

Figure 1 shows the developed optical diagnostic system for spray visualisation [20]. The laser beam was generated by a double-pulse Nd:YAG green laser (532 nm wavelength), and then turned into a thin laser sheet (around 3 mm in width at the imaging plane). The maximum optical power of the laser device was 600 mJ during a pulse-width as short as 4 ns. Two CCD cameras (Table.1) acquired the images at both macroscopic and microscopic levels. A controller synchronised the injection pulses with the laser pulses and the camera exposure using a programmable hardware with a TI-MSP430 microcontroller core and an embedded-C program. A Scheimpflug adaptor was mounted between the lens and the photo-sensor to keep the object plane is in sharp focus.

The cameras produced double-frame images for particle image velocimetry PIV tests. The time interval was 12-50 μs between the frame couples and 25-50 μs between the test points (depending on the spray maximum velocity). The result of each test point was represented by the average of 250-500 images, compared to 5-50 images reported in most similar experiments.

This system was previously employed to calculate the tip penetration velocity and the microscopic velocity using micro-PIV for a standard gasoline injector (D_o=584 μm) [20]. In the current paper, the results are compared against another injector.

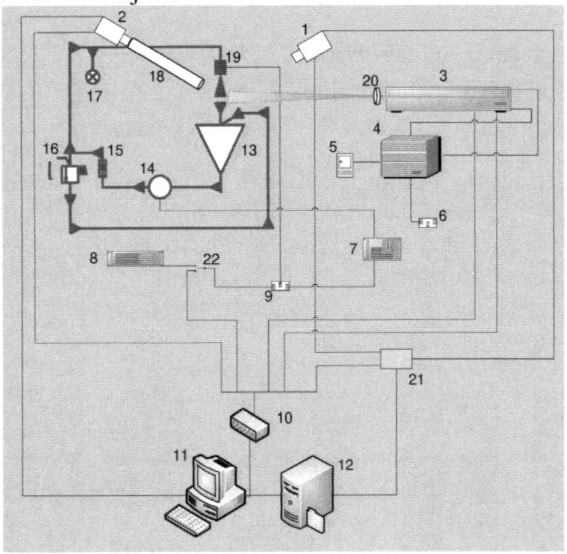

Figure 1. The spray visualisation rig for high-speed spray jet: (1) high resolution first CCD camera, (2) second CCD camera, (3) Nd:YAG laser head, (4) laser power supply, (5) laser controller, (6) 3-Amps AC main power, (7) adjustable DC power supply, (8) TTL signal generator, (9) injector driver circuit, (10) PC-control unit, (11) control PC, (12) data storage RAID, (13) fuel supply, (14) fuel pump, (15) flow control, (16) pressure relief valve, (17) pressure gauge, (18) K2 microscope, (19) fuel Injector, (20) laser sheet optics, (21) PCO camera controller, (22) trigger switch.

Table 1. Cameras specifications in the spray visualisation system.

	PCO Pixelfly	PCO 2000
Resolution (pixel)	1360 x 1024	2048 x 2048
Frame-rate (Hz, Max)	50	14.7
Inter-frame time (ns)	5000	180
Dynamic Range (bits)	12	14
Exposure (μs, Min)	5	0.5

4. Results and discussion

The mean-image was calculated at each test point. The mean-images were then binarized by a threshold that separated the spray pixels from the background. Due to the low density of the sprays, it was necessary to fill up the gaps within the spray image using a filter (3x3 mask) that compares each pixel with its neighboring pixels. Finally an edge detection procedure defined the spray boundaries using a 2D-median filter. Figure 2 shows the steps of acquiring the spray profile of the micro dispensing valve using a series of injection pulses (416.7 Hz, 50% duty-cycle). The resulting images directly provided the cone-angle and the sheet area.

We also analysed spray images at several stages of the injection cycle to observe the fluid disintegration process step by step. Figure 3 shows example images using a standard gasoline injector at 5-bar line-pressure. The injector valve was opened for 1.2 ms and closed for another 1.2 ms. The sequenced images show that the primary atomisation started 300 μs ASOI. At 450 ASOI the fluid bulk started disintegrating off the jet surface. The disintegration process continued as the jet penetrated further downstream of the nozzle. By the end of the injection pulse, it could be seen that the small droplets produced by the early atomisation stage stayed behind and moved much slower than the central jet. The breakup of the central jet happened after the end of the injection pulse, producing only large droplets. The average penetration by the end of the injection pulse was approximately 38 mm for the 5 bar line-pressure, compared to 20 mm using 3 bar line-pressure.

In this part of the experiment, the focus was on the central jet progression using a standard (55 mm) lens. Droplets that are less than 50 μm in diameter were visible to the microscope only. This is because the diffraction efficiency in the case of large fluid drops is much higher than in the small droplets at the 90° angle of observation [21]. The flow-rate distribution measurements (using mechanical patternators) showed that for the gasoline injector almost 40% of the injected fluid stays at the central part of the spray (less than 20° divergence angle), compared to 28% for the micro dispensing valve. The central part of the spray represents the poorly atomised liquid, while the fine droplets are drifted away towards the peripheries.

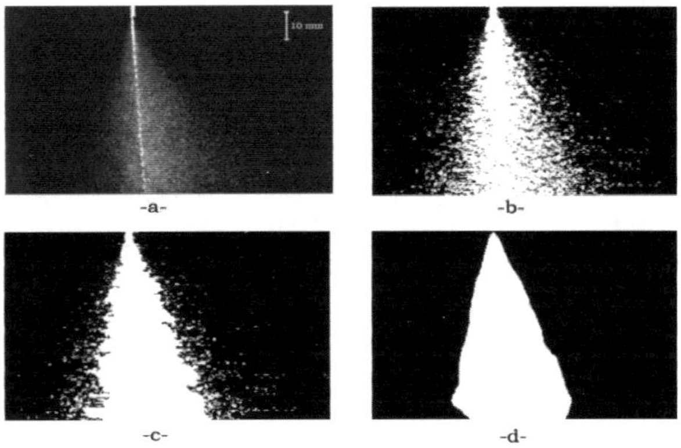

Figure 2. Spray boundaries detection by image processing of a micro dispensing valve at 3 bar: (a) mean-image, (b) black and white image, (c) gaps filled up, (d) spray boundaries by edge detection. The angle between the laser sheet and the camera sensor was 60° (focus was corrected by a Scheimpflug adaptor).

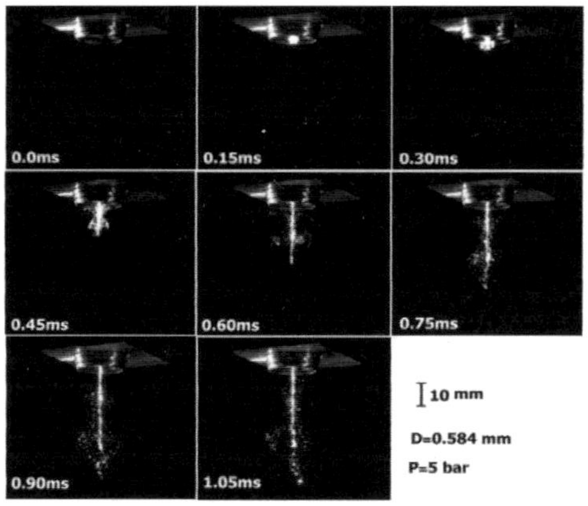

Figure 3. Images of the JP8 spray during an injection pulse of 1.2 ms using a gasoline injector (D_o=584 μm). Images were taken for several injection incidents using a 2 mm laser sheet (4 ns optical pulse-width). The angle between the laser sheet and the camera sensor was 90°.

The tip penetration measurements (Figure 4) show that sprays penetrated around 7 cm by the end of the injection pulse for both nozzles at Pin =5 bar, compared to around 3 cm at Pin =4 bar. The discharge velocity was lower in the micro nozzle sprays than in the gasoline injector. The tip velocity of the micro nozzle sprays dropped at higher rates than the gasoline injector sprays, which was an indication for a better atomisation at the centre of the jet. Figure 5 shows the tip velocity calculated from the tip penetration results by calculating the first derivative of the third order polynomial fit of the graphs in Figure 4. The tip velocity decreased at higher rates for sprays produced by high pressure levels.

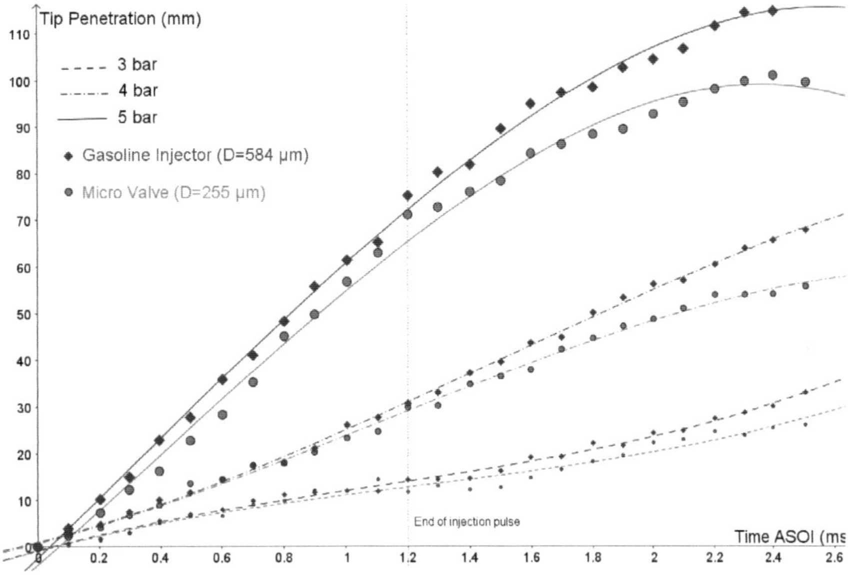

Figure 4. Tip penetration for JP8 sprays against time ASOI (P_{in} = 3,4,5 bar) for the two tested nozzles.

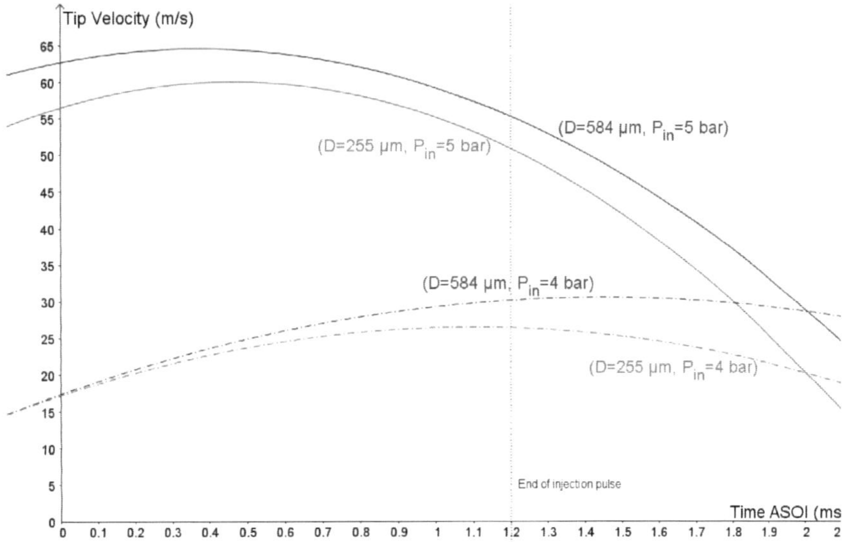

Figure 5. Tip velocity for JP8 sprays against time ASOI (P_{in} = 4,5 bar) estimated from the tip penetration data.

The colour coded mean-image of the fully developed spray (Figure 6) provides a statistical prospective of the spray density distribution. The middle highly dense region of the fluid jet penetrated far downstream of the injector's nozzle, while low droplet density with lower velocity elements (due to the smaller mean droplet diameter) scattered to wider angles. Although the gasoline injector had a larger discharge angle closer to the orifice (due to its larger orifice diameter), most of the high density elements were concentrated in the middle of the spray jet. In contrast, the middle of the spray produced by the micro dispensing nozzle disintegrated into smaller elements that diverged into wider angles further downstream. The density maps provide a good representation of how the fluid volume is distributed throughout the spray area. The local velocity for these elements was measured by a particle image velocimetry (2D- PIV) analysis.

PIV is a well-established non-intrusive measurement technique in fluid mechanics, where the velocity vector field is generated by calculating the displacement of the particles between two successive frames. Each frame is divided into fixed size interrogation windows. Each window from the first frame is compared by the corresponding window from the second frame by calculating the mathematical cross-correlation. The velocity vector field (Figure 7) shows that at the middle jet velocity was 30-35 m/s at the end of the injection pulse. This is comparable to what we have found in the tip velocity graphs (Figure 5), where 30 m/s was estimated for the gasoline injector and 26.6 m/s for the micro dispensing system at the same instant ASOI. It is possible to see a higher concentration of the high velocity elements at the middle and the tip regions in the gasoline injector case. It shows that the micro dispensing nozzle performs earlier atomisation than the gasoline injector, and this is especially useful in the event of engine's cold start.

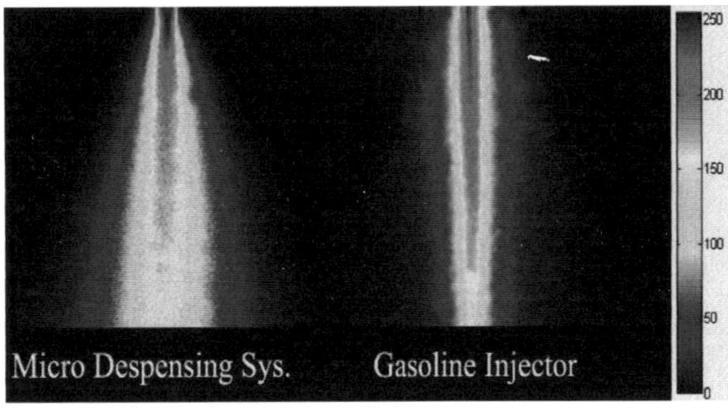

Figure 6. The fluid density distribution represented by the average pixel intensity of spray images for a standard gasoline injector and a micro dispensing nozzle (P_{in}= 4 bar, T= 1.2 ms ASOI, continuous injection).

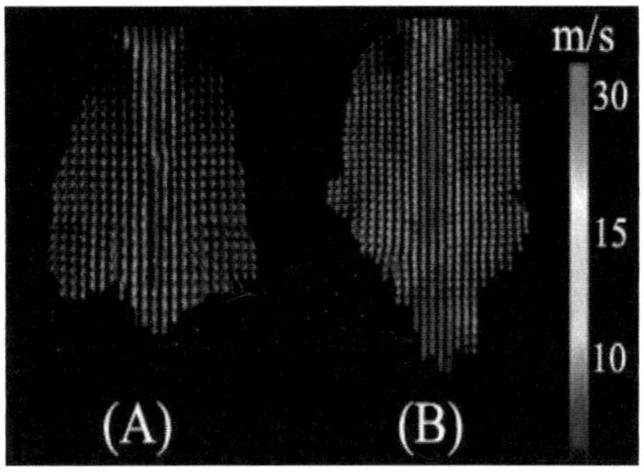

Figure 7. Velocity vector fields of JP8 fuel sprays (T=1.2 ms ASOI, P_{in}= 4 bar) by two-stage cross-correlation for (A) a micro dispensing nozzle, (B) a gasoline injector.

5. Conclusions

We studied JP8 fuel sprays produced by two fluid atomisers. The first atomiser was a standard gasoline injector for low-flow rate applications. The second was a light-weight high-frequency micro-dispensing system. The fuel sprays were visualised using a laser-sheet imaging technique and acquired the macroscopic characteristics of the sprays with image processing. Understanding the development process of fuel sprays is important when designing engines, especially small rotary engines, where time available for atomisation, vaporization and mixing with air is relatively short.

The discharge velocity was lower in the micro dispensing nozzle sprays than in the gasoline injector. However, the tip velocity of the micro nozzle sprays dropped at higher rates for the micro dispensing nozzles and for high pressure rates. The micro dispensing nozzle seemed to perform an earlier atomisation than the gasoline injector, producing low-speed small droplets that scatter at wider angles. The PIV test confirmed that the standard gasoline injector produced sprays with higher velocity at the centre and the tip regions. These results are to be confirmed in the future by droplet size and velocity distribution analysis using long-range microscopy.

Acknowledgments
The authors would like to thank the Optical Engineering Laboratory (OEL) team at the University of Warwick, and Cubewano Ltd R&D team for their support of the work presented in this paper. This work was financially supported by Hansung University.

References

[1] Crua C, Heikal M R and Gold M R 2015 Microscopic imaging of the initial stage of diesel spray formation *Fuel* **157** pp 140-50
[2] Dong Q, Ishima T, Kawashima H and Long W 2013 A study on the spray characteristics of a piezo pintle-type injector for DI gasoline engines *Journal of Mechanical Science and Technology* **27**(7) pp 1981-93
[3] Costa M, Sorge M and Allocca L 2012 Increasing energy efficiency of a gasoline direct injection engine through optimal synchronization of single or double injection strategies, *Energy Conversion and Management* **60** pp 77–86
[4] Hiroyasu H and Arai M 1990 Structures of fuels sprays in Diesel engines *SAE technical paper* 900475
[5] Payri R, Viera J P, Gopalakrishnan V and Szymkowicz P G 2017 The effect of nozzle geometry over the evaporative spray formation for three different fuels *Fuel* **188** pp 645-60
[6] Payri R, Viera J P, Gopalakrishnan V and Szymkowicz P G, 2016 The effect of nozzle geometry over internal flow and spray formation for three different fuels *Fuel* **183** pp 20–33
[7] Lad V N and Murthy Z V P 2016 Effects of the geometric orientations of the nozzle exit on the breakup of free liquid jet *Journal of Mechanical Science and Technology* **30** (4) pp 1625-30
[8] Yoon S H, Kim D Y, Kim D K and Kim B H 2011 Effect of nozzle geometry for swirl type twin-fluid water mist nozzle on the spray characteristic, *Journal of Mechanical Science and Technology* **25** (7) pp 1761-66
[9] Liu Z, Im K S, Wang Y, Fezzaa K, Wang J, Xie X B, Lai M C and Wang J 2010 Near-nozzle structure of diesel sprays affected by internal geometry of injector nozzle: visualized by single-shot X-ray imaging *SAE technical paper* 2010-01-0877
[10] Han J S, Lu P H, Xie X B, Lai M C and Henein N A 2002 Investigation of diesel spray primary break-up and development for different nozzle geometries, *SAE technical paper* 2002-01-27
[11] Chen P C, Wang W C, Roberts W L and Fang T 2013 Spray and atomization of diesel fuel and its alternatives from a single-hole injector using a common rail fuel injection system *Fuel* **103** pp 850-61
[12] Delacourt E, Desmet B and Besson B 2005 Characterisation of very high pressure diesel sprays using digital imaging techniques *Fuel* **84** (7) pp 859-67
[13] Suh H K and Lee C S 2008 Effect of cavitation in nozzle orifice on the diesel fuel atomization characteristics *Int. J. Heat Fluid Flow* **29** (4) pp 1001–09
[14] Blessing M, König G, Krüger C, Michels U and Schwarz V 2003 Analysis of flow and cavitation phenomena in diesel injection nozzles and its effects on spray and mixture formation *SAE technical paper* 2003-01-1358
[15] Naber J D and Siebers D L 1996 Effects of Gas Density and Vaporization on Penetration and Dispersion of Diesel Sprays *Transactions of the SAE* 960034
[16] Harris A S, Svensson E, Wagner Z G, Lethagen S and Nilsson I M 1988 Effect of viscosity on particle size, deposition, and clearance of nasal delivery systems containing desmopressin *Journal of pharmaceutical sciences* **77** (5) pp 405-408
[17] Arai M, Tabata M, Hiroyasu H and Shimizu M 1984 Disintegrating process and spray characterization of fuel jet injected by a diesel nozzle *SAE Technical Paper* 840275
[18] Lefebvre A 1988 *Atomization and sprays* (CRC press) vol 1040 no 2756
[19] Payri R, Garcia-Oliver J M, Bardi M and Manin J 2012 Fuel temperature influence on diesel sprays in inert and reacting conditions *Applied Thermal Engineering* **35** pp 185-95
[20] Zakaria R, Bryanston-Cross P and Timmerman B 2014 Spray development process of aviation fuel using a low-pressure fuel injector: Visualization and analysis *Journal of Mechanical Science and Technology* **28** (12) pp 5003-11
[21] Zakaria R and Bryanston-Cross P 2012 Light scattering efficiency of oil smoke seeding droplets in PIV systems *IEEE Photonics Global Conference (PGC)* pp 1-5

IOP Conf. Series: Materials Science and Engineering **282** (2017) 012002 doi:10.1088/1757-899X/282/1/012002

Computer-aided solvent selection for multiple scenarios operation of limited-known properties solute

Amata Anantpinijwatna

Department of Chemical Engineering, Faculty of Engineering, King Mongkut's Institute of Technology Ladkrabang, Bangkok, 10520, Thailand

*E-mail: amata.an@kmitl.ac.th

Abstract. Solvents have been applied for both production and separation of the complex chemical substance such as the pyrrolidine-2-carbonyl chloride (C_5H_8ClNO). Since the properties of the target substance itself are largely unknown, the selection of the solvent is limited by experiment only. However, the reaction carried out in conventional solvents are either afforded low yields or obtained slow reaction rates. Moreover, the solvents are also highly toxic and environmental unfriendly. Alternative solvents are required to enhance the production and lessen the harmful effect toward both organism and environment. A costly, time-consuming, and laborious experiments are required for acquiring a better solvent suite for production and separation of these complex compounds; whereas, a limited improvement can be obtained. On the other hand, the combination of the state-of-the-art thermodynamic models can provide faster and more robust solutions to this solvent selection problem. In this work, a framework for solvents selection in complex chemical production process is presented. The framework combines a group-contribution thermodynamic model and a segment activity coefficient model for predicting chemical properties and solubilities of the target chemical in newly formulated solvents. A guideline for solvent selection is also included. The potential of the selected solvents is then analysed and verified. The improvement toward the production yield, production rate, and product separation is then discussed.

1. Introduction

Multiphase reacting system has broad range of application area including the manufacture of petroleum-based products, the production of chemicals, pharmaceuticals, and agro-bio products. Some important examples include: (1) phase transfer catalyst (PTC) involved systems, which has been recognized since 1940s, receiving much attention as an efficient alternative organic synthesis operation [1,2]; (2) epoxidation of unsaturated fatty acid by hydrogen peroxide to produce phthalates substitution [3,4]; (3) hydroformylation processes that produce aldehydes from olefins [5]; (4) production of pyrethroid compound insecticide [6,7]; and, (5) biodiesel production [8]. In these systems, multiple phases are created by the immiscibility between co-exist solvents such as aqueous/organic, ammonium/organic, supercritical-CO2/ionic liquid, or separated by permeable membrane. The reactants and catalysts (including also biocatalysts and enzymes) can exist in different liquid phases, allowing novel synthesis paths as well as higher selectivity, conversion, and yield. Furthermore, after the reaction, reactants, catalyst and products may end up in different liquid phases, making the separation tasks easier.

Phenomena related to this system including partitions and diffusions of each chemical species between phases, dissociation of inorganic salt in aqueous phase, reactions, and accumulation of chemical species at interface has already been described mathematically in the previous work [9,10]. Solvent selection is one of the main aspect toward the reactor design since it could help speeding up the reaction; as well as, reducing the separation effort. However, the effect of the solvents toward the production and separation were barely described. Moreover, due to the complexity of the involved chemical substances, the required properties for the calculation are largely unknown.

In this work, systematic framework for solvent selection is formulated and presented. Combining with the multiphase reacting model constructed by the previous work, it could help accelerating the reaction; as well as minimizing the separation effort to produce pyrrolidine-2-carbonyl chloride (PCC, C_5H_8ClNO) process.

2. Solvent selection methodology

The methodology for selecting good solvents for the reaction is presented in figure 1. The methodology is divided into 3 parts: solute properties estimation, solvent usability estimation, and solvents collection and generations.

2.1. Solute properties estimation

To choose a good solvent, first the properties of solute must be estimated first. The solute properties, such as melting temperature and heat of fusion, are collected from the database; or, if not available, estimated by the well-described group-contribution method [11,12].

2.2. Solvent usability test

The solubilities and the partitions of the desired solute in various solvents, distinguished by solubility parameters [13], are then calculated. Due to the limit in availability of the binary interaction between groups of the group-contribution based method such as UNIFAC, these values are calculated from the simpler segment-based thermodynamic model (NRTL-SAC, [14,15]). The preferred solvent is then identified.

2.3. Novel solvents collection and generation

The solvent with similar properties are then collected and verified. By using pure compound properties as screening criteria, novel solvents can also be generated through group contribution based method.

The selected solvents from this methodology are then used as substitute solvent for calculating the activity coefficients, partitions, rates, and balance in the model in the next section. In this work, the conventional solvents and novel solvents, obtained from this methodology, of the production of PCC process are employed to demonstrate the capability of this methodology.

3. Model of the PCC production process

Pyrrolidine-2-carbonyl chloride (PCC) is produced from the reaction between proline and oxalyl chloride in the presence of dimethyl formamide (DMF) catalyst as shown in reaction 1.

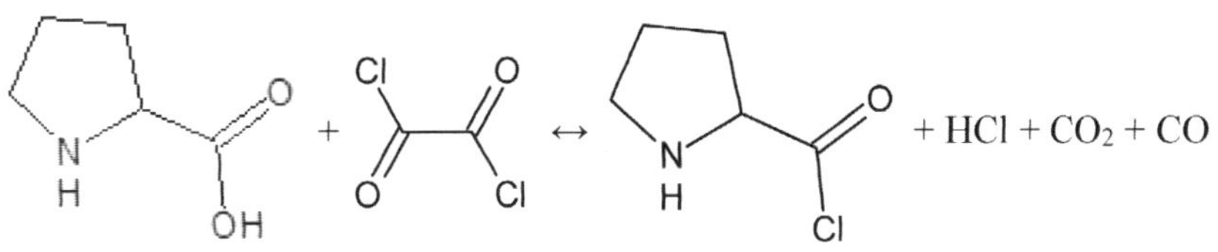

(1)

(Proline) (Oxalyl Chloride) (PCC)

The reaction is carried out in biphasic water-solvent systems with less than 40% conversions in the conventional solvents: dichloromethane, chloroform, chlorobenzene, and dioxane. Moreover, the PCC product and proline can also further be dimerized into unwanted amino product as shown in reaction 2.

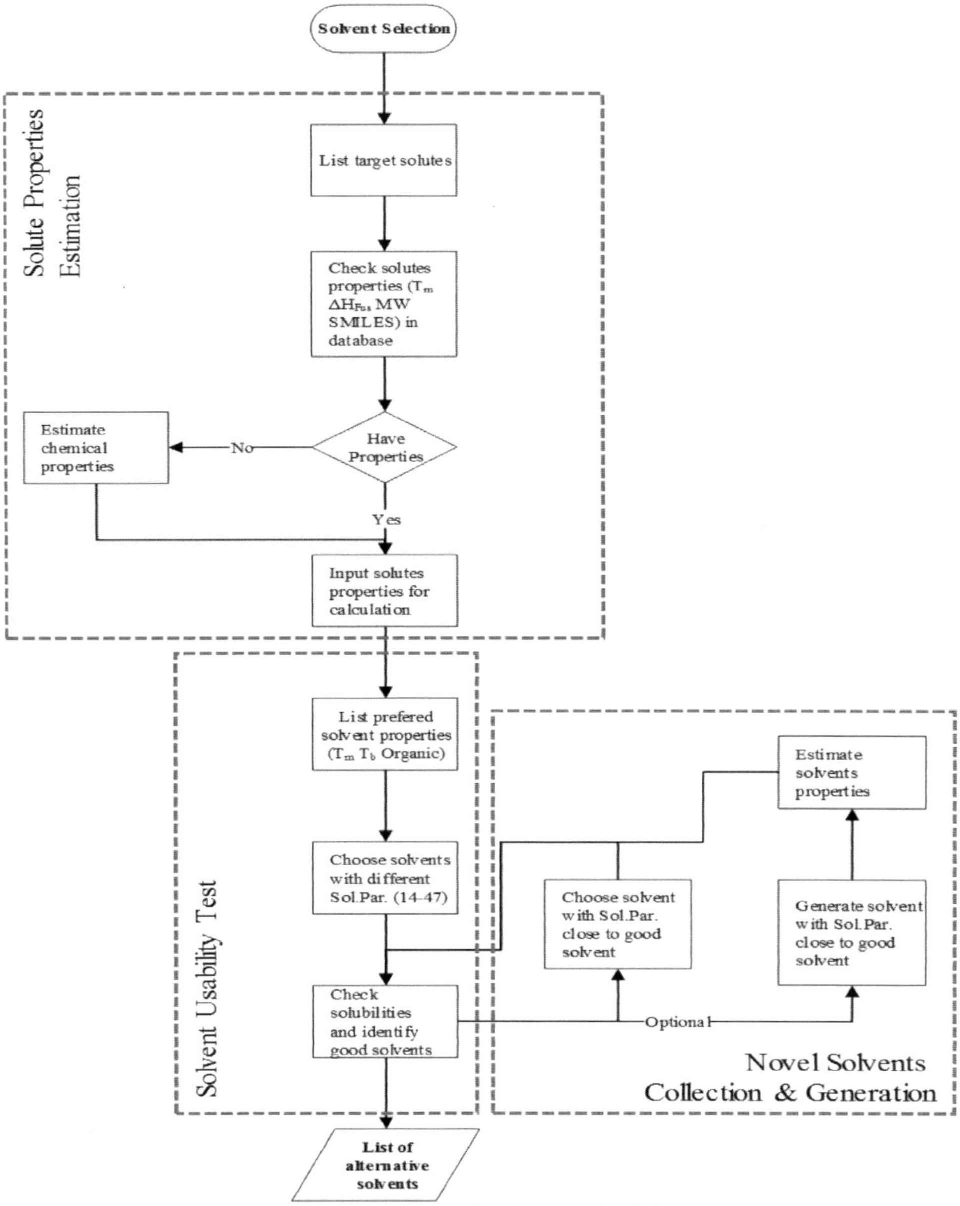

Figure 1. Solvent Selection Methodology.

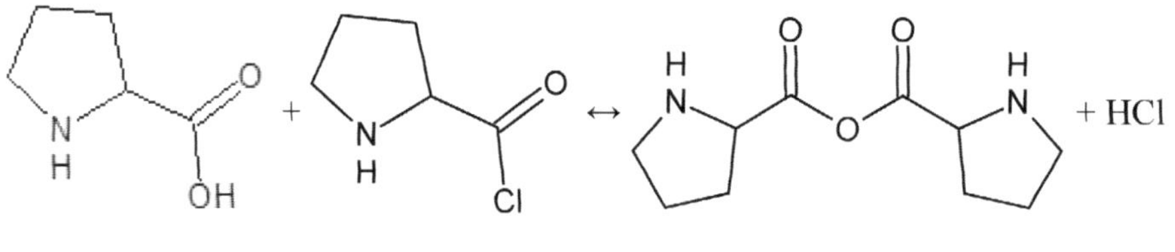 (2)

The mathematical model describes this reacting system is constructed with the multiphase modelling framework in my previous work (9]. PPC, proline, oxalyl chloride, and the DMF catalyst are considered to be heterogeneous species, while the other chemicals are considered to be homogeneous compound. Hence, from module 1 of the framework, the partitions of the heterogeneous compounds between the aqueous and the organic phases are described by the following equations 3-6.

$$P_P = \frac{\gamma_P^\alpha}{\gamma_P^\beta} \tag{3}$$

$$P_O = \frac{\gamma_O^\alpha}{\gamma_O^\beta} \tag{4}$$

$$P_{PC} = \frac{\gamma_{PC}^\alpha}{\gamma_{PC}^\beta} \tag{5}$$

$$P_D = \frac{\gamma_D^\alpha}{\gamma_D^\beta} \tag{6}$$

where P_i denotes the partition of the i compound, γ_i^j denotes the activity coefficient of the i compound in the j phase which are calculated from the designated thermodynamic model(s), α and β denote the phase aqueous and organic, and *P, O, PC, D* denote the compounds proline, oxalyl chloride, PCC, and DMF respectively.

The reversible reactions are only taking place in the organic phase; hence, according to the module 2 of the framework, the reaction rates are expressed as the following equations 7 and 8.

$$R_1 = k_1\left(C_P^\beta \gamma_P^\beta C_O^\beta \gamma_O^\beta - \frac{C_{PC}^\beta \gamma_{PC}^\beta}{K_1^{Eq}}\right) \tag{7}$$

$$R_2 = k_2\left(C_P^\beta \gamma_P^\beta C_{PC}^\beta \gamma_{PC}^\beta - \frac{C_W^\beta \gamma_W^\beta}{K_1^{Eq}}\right) \tag{8}$$

where C_i^j denotes the concentration of the species i in the phase j, W denotes the waste species, R_m denotes the rate of reaction, k_m and K_m^{Eq} denote the rate coefficient, and the equilibrium constant of reaction m respectively.

From the module 3 of the framework, the balance equations of the reactive species are established as following equations 9-17, assuming the reactor is a continuous fed-batch reactor.

$$\frac{d\xi_1}{dt} = R_1 V^\beta \tag{9}$$

$$\frac{d\xi_2}{dt} = R_2 V^\beta \tag{10}$$

$$N_P^\alpha = N_P^0 - \xi_1 - \xi_2 + F_P^0 t - N_P^\beta \tag{11}$$

$$N_O^\alpha = N_O^0 - \xi_1 + F_O^0 t - N_O^\beta \tag{12}$$

$$N_{PC}^\alpha = N_{PC}^0 + \xi_1 - \xi_2 + N_{PC}^\beta \tag{13}$$

$$N_{HCl}^\alpha = N_{HCl}^0 + \xi_1 + \xi_2 \tag{14}$$

$$N_{CO_2}^{\alpha} = N_{CO_2}^{0} + \xi_1 \quad (15)$$

$$N_{CO}^{\alpha} = N_{CO}^{0} + \xi_1 \quad (16)$$

$$N_{W}^{\beta} = N_{W}^{0} + \xi_2 \quad (17)$$

where ξ_m denotes the extent of the reaction *m*, V^j denote the volume of the phase *j*, N_i^0 denotes the initial amount of the species *i*, *t* denotes the reaction time, F_i^0 denotes the inlet flow of the species *i*, and N_i^j denotes the amount of the species *i* in the phase *j* at any given time *t*.

4. Application and results

For the PCC production process, the substitute solvents should be liquid at operating condition (boil at temperature more than 350 K), they should help improving the solubilities of the reactant, accelerating the reaction, and purifying the product.

First, the reactants (proline and oxalyl chloride) and product (PCC) are considered to be target solutes; their properties are estimated by group contribution method [16] and shown in table 1. Heat of fusion (ΔH_{Fus}) and melting temperature (T_m) are required for solubilities calculations, Hansen and Hildebrand solubility parameters [17,18] are used for solvent screening.

Table 1. Estimated target properties of solutes.

Chemical	Sol Par.	δD	δP	δH	ΔH_{Fus} (kJ/mol)	T_m (K)	T_b (K)
Proline	24.83	17.54	6.28	12.33	15.40	358.69	521.66
Oxalyl Chloride	16.54	16.34	7.69	6.99	12.53	262.99	394.05
PCC	19.1	24.46	12.41	20.08	18.03	315.38	471.68

The solvents with different solubility parameters are then tested with proline and PPC for their solubilities, the results are shown in figures 2 and 3 respectively. It is appeared both proline and PCC are well-soluble in the solvents which have solubility parameters between 26-32; therefore, the solubility parameters limit are included into the criteria for solvent selection.

The solvents are then created and screened according to the criteria listed in table 2. The desired solvents should be liquid at room temperature, have boiling point match the operating condition, and shouldn't contain ethers or esters group to prevent the reaction with proline.

From these criteria, more than 150 thousand solvents are created; hundreds of solvents pass the criteria; and 52 solvents are the common solvents in the database. These solvents are then classified for different usage based on the operation scenarios and calculation from the model from section 3. Only solvents that provide noticeable improved results will be displayed and discussed.

Table 2. Solvents screening criteria.

	Constraint
T_b	>350
T_m	<293
Sol Par.	26-32
δ_D	14-20
δ_P	3-10
δ_H	2-9
Group	No ethers and esters

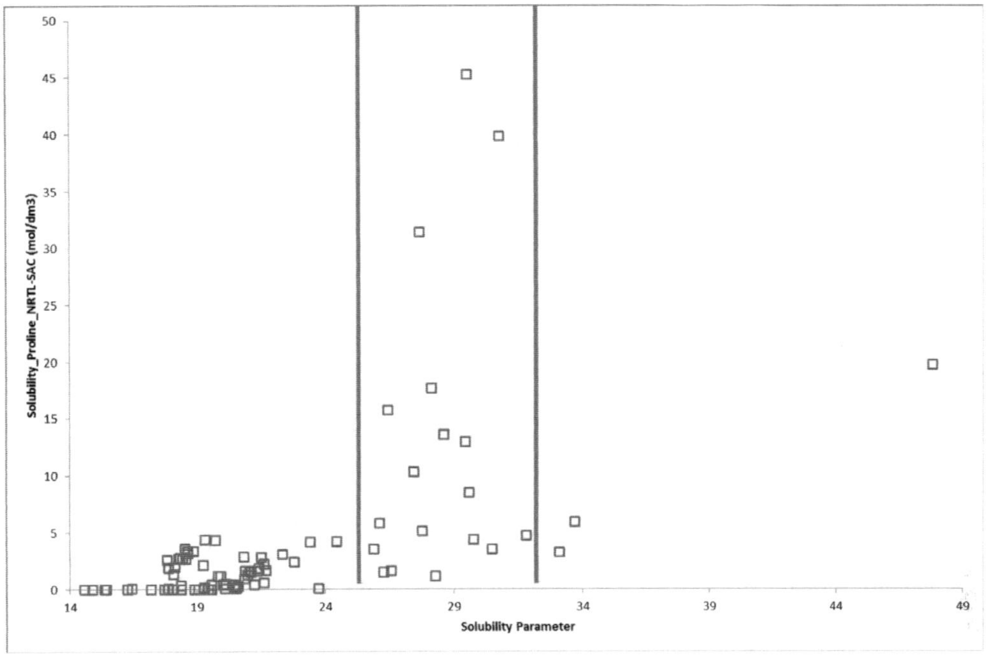

Figure 2. Solubilities of proline in different solvent.

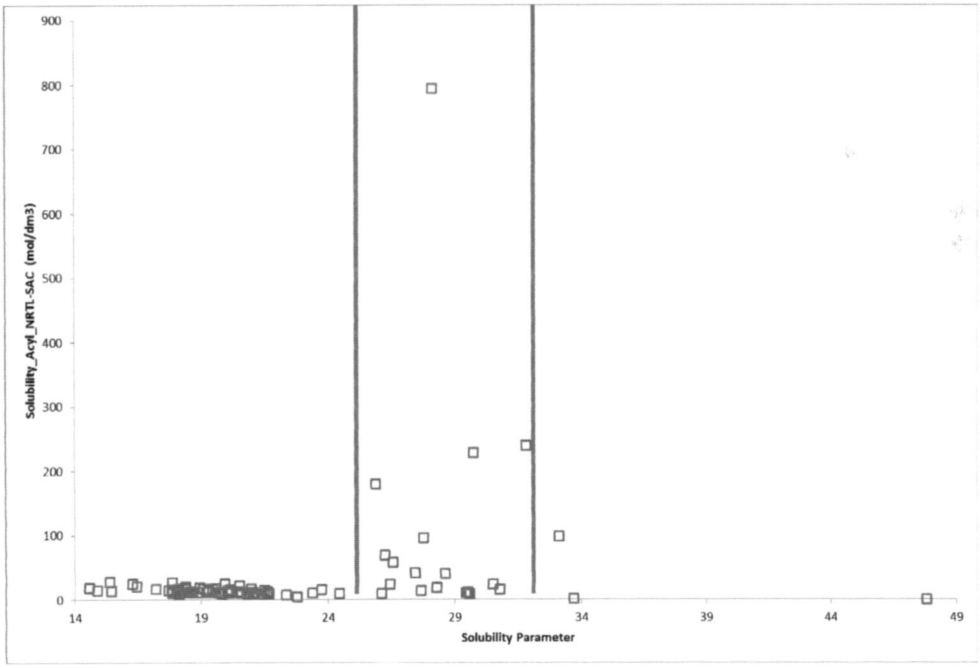

Figure 3. Solubilities of PCC in different solvent.

4.1 Single solvent operation

The single solvent operation assumed using a solvent with high solubility for both proline and PCC for simplest operation, the solvent work as carrier for reactants and product; as well as, reactive phase. The solvents, their properties, and the improvement on the rates of reaction are shown in table 3.

Table 3. Selected solvents for single solvent operation scenario.

Solvent	Sol Par.	T_m	T_b	Solubilities (mol/dm^3)		Improvement (Time)
				Proline	PPC	
Methyl-diethanolamine	28.14	252.2	518.0	17.69	794.37	3.78
Hydracrylonitrile	28.61	227.2	494.2	13.59	39.61	11.60
3-amino-1-propanol	29.76	284.2	460.7	4.39	227.76	1.11
N-methylformamide	30.50	269.4	472.7	3.53	23.79	1.81

All selected solvents improve the solubilities of both proline and PPC; however, only hydracrylonitrile gives extensive improvement on the rate of reaction (11.60 time compare to conventional solvent). Because of the reversible reaction from PPC back to proline and side reaction between proline and PPC, the solvents with high PPC solubilities provide limited improvement.

4.2 Single phase, multiple solvents

In this scenario, 2 solvents are combined, the first solvent is fed together with proline-oxalyl chloride reactants, where the reaction is taking place in this solvent. The second solvent, which is completely miscible with the first solvent, is then fed into the system for dissolving PCC product. The selected solvents, their solubility parameters, and the improvements are reported in table 4.

For both 1st solvents the rates of reaction are increasing due to the increasing solubility of proline; the reaction in 1,3-propylene-glycol has a bit higher rate of reaction because of the solubility of proline is higher than in 2-methyl-1,3-propanediol; however, the 2nd solvents have similar trend of impact toward the reaction rates.

Table 4. Selected solvents for single phase, multiple solvents operation scenario.

Solvent 1	Sol Par.	Solvent 2	Sol Par.	Improvement (Time)
2-methyl-1,3-propanediol	27.68	**Methyl-diethanolamine**	28.14	18.54
		Monoethanolamine	31.83	17.35
		3-amino-1-propanol	29.76	19.21
		1-amino-2-propanol	25.89	22.50
1,3-propylene-glycol	30.78	**Methyl-diethanolamine**	28.14	24.51
		Monoethanolamine	31.83	21.23
		3-amino-1-propanol	29.76	25.51
		1-amino-2-propanol	25.89	30.33

4.3 Multiphase, multiple solvents

The last scenario operates using 2 solvents; the first solvent is fed together with proline-oxalyl chloride reactants, where the reaction is taking place in this solvent; however, in this scenario, the second solvent is not miscible with the first solvent and it sole purpose is to extract the PCC product from the reacting phase, hence accelerating the reaction further. The selected solvents, their solubility parameters, and the improvements are reported in table 5.

In this scenario, the improvements are more extensive compare to other scenario; moreover, the final product is extracted from other impurities making it easier for separation steps. Nonetheless, the design of the reactor and the complexity reactor operation must be taken into account; the revised design-operation of the reactor is highly advised.

Table 5. Selected solvents for multiphase, multiple solvents operation scenario.

Solvent 1	Sol Par.	Solvent 2	Sol Par.	Improvement (Time)
Water	47.81	**Beta-Propiolactone**	26.58	40.23
		2-Methylnonane	15.38	55.57
		2,6-Dimethyl-4-Heptanol	17.87	57.80
		Isopropylcyclohexane	16.29	56.79
Allylamine	19.93	**2-methyl-1,3-propanediol**	27.68	54.55
		1,2-propylene-glycol	25.89	53.57
		1,3-propylene-glycol	30.78	55.54

5. Conclusion

A solvent selection methodology help systematizing the solvent selection procedure. The combination of group contribution and segment contribution method help broadening the range of application of the system. This methodology has been applied to the complex pyrrolidine-2-carbonyl chloride production process. Together with the model built from the multiphase modelling framework, improvement toward production rate can be achieved.

References

[1] W H C Rueggeberg, A Ginsbu'rg and R K Frantz 1946 *Ind. Eng. Chem.* **38** pp 207–11
[2] S Shirakawa and K Maruoka 2013 *Angew. Chem. Int. Ed. Engl.* **52** pp 4312–48
[3] E Santacesaria, G Vicente, M Di Serio and R Tesser 2012 *Catal. Today* **195** pp 2–13
[4] L H Gan, S H Goh, K S Ooi and J Am 1992 *Oil Chem. Soc.* **69** pp 347–51
[5] H Nowothnick, A Rost, T Hamerla, R Schomäcker, C Müller and D Vogt 2013 *Catal. Sci. Technol.* **3** pp 600–05
[6] A A Chrisochoou, K Schaber and K Stephan, 1997 *J. Chem. Eng. Data* **42** pp 551–57.
[7] S Panke, M Held and M Wubbolts 2004 *Curr. Opin. Biotechnol.* **15** pp 272–9.
[8] H Amiri, K Karimi and S Roodpeyma 2010 *Carbohydr. Res.* **345** pp 2133–8.
[9] A Anantpinijwatna, S H Kim, M Sales-Cruz, J P O'Connell and R Gani 2016 *J. Chem. Eng. Data* **61** pp 4090–103
[10] A Anantpinijwatna, M Sales-Cruz, S H Kim, J P O'Connell and R Gani 2016 *Chem. Eng. Res. Des.*
[11] J Marrero and R Gani 2001 *Fluid Phase Equilib.* pp 183–208
[12] A T Karunanithi, L E K Achenie and R Gani 2005 *Ind. Eng. Chem. Res.* **44** pp 4785–97
[13] C M Hansen 1969 *Ind. Eng. Chem. Prod. Res. Dev.* **8** pp 2–11
[14] C Chen and P Crafts 2006 *Ind. Eng. Chem. Res.* pp 4816–24
[15] C -C Chen and Y Song 2004 *Ind. Eng. Chem. Res.* **43** pp 8354–62
[16] H E González, J Abildskov and R Gani 2007 *Fluid Phase Equilib.* **261** pp 199–204
[17] J B Durkee 2004 *Met. Finish.* **102** pp 42–50
[18] C M Hansen *2007 Hansen Solubility Parameters: A User's Handbook, Second Edition, CRC Press*

The performance of double layer structure membrane prepared from flowing coagulant

Chan Mieow Kee[1,*], Anthony Leong Chan Xeng [2], Sasiskala Regal[1], Balvinder Singh[1], Preeshaath Raoo[1], Yap Koon Eu[1] and Ng Sok Choo[3]

[1]Faculty of Engineering and the Built Environment, SEGi University. 9, Jalan Teknologi, Taman Sains Selangor, PJU 5, Kota Damansara, 47810 Petaling Jaya, Selangor, Malaysia
[2]Taylor's College. 1, Jalan SS15/8, SS15, 47500 Subang Jaya, Selangor, Malaysia
[3]School of Information Technology, SEGi University, Malaysia

*E-mail: mkchan@segi.edu.my

Abstract. Membrane with double layer structure is favourable as it exhibits smooth surface and macrovoids free structure. However, its' performance in terms of permeability, porosity and strength has not been studied thoroughly. Additionally, the effect of flowing coagulant on the formation of double layer membrane has not been reported. Thus, the objective of this study is to investigate the performance of double layer membranes, which were prepared using flowing coagulant. Results showed that when the coagulant flow changed from laminar to turbulent, the pure water permeation of the membrane increased. It was due to the higher porosity in the membrane, which prepared by turbulent flow (CA-Turbulent) compared to the membrane which fabricated under laminar condition (CA-Laminar). This can be explained by the rapid solvent-coagulant exchange rate between the polymer solution and the turbulent coagulant. In term of strength, the tensile strength of the CA-Turbulent was ~32 MPa, which was 100% higher compared to CA-Laminar. This may due to the presence of large amount of nodules on its surface, which reduced the surface integrity. In conclusion, flowing coagulant altered the membrane properties and adopting turbulent coagulant flow in membrane fabrication would improve the porosity, surface roughness and the strength of the membrane.

1. Introduction

Polymeric membranes are widely used for separation purpose due to low cost [1], well developed module technology and available in varied materials [2]. However, polymers such as polyethersulfone (PES) and polyvinylidene fluoride (PVDF) are hydrophobic in nature, which exhibit high fouling tendency. Natural polymer like cellulose acetate (CA) is weak in mechanical strength which limits its usage in high pressure operating condition. Thus, efforts have been done to improve the membrane properties by adding new additives into the formulation [3-4], introducing new fabrication technique [5] and surface modification [6-7].

Double pass casting technique was first introduced by Dang et al. [8] and this technique improved the surface roughness of hydrophobic PES membrane. Besides, the researchers also observed macrovoids free structure in the modified hydrophilic PES membrane. However, this fabrication technique has not been studied extensively by the researchers, especially on different types of material.

The effect of the flowrate of bore fluid in fabricating hollow fibres membrane has been widely studied by the researchers. The finding from Alsalhy et al. [9] showed that high bore fluid rate produced PES membranes with high water flux property by altering the pore size and the porosity of the membranes. Similar finding also reported by Qin et al. [10] where the flowrate of the bore fluid affected the selectivity of gas membrane. In flat sheet membrane fabrication, the coagulant plays the same role as the bore fluid in introducing the phase inversion process. Nevertheless, the effect of flowing coagulant on the performance of flat sheet membranes has not be studied.

Therefore, the objective of this study is to bridge the gap in the existing research literature by investigating the performance of double layer structured CA membranes, which were fabricated using flowing coagulant. The membranes were evaluated in terms of water permeation, hydrophilicity and tensile strength.

2. Materials

CA (average molecular weight, 100,000, ACROS Organics™) was used as the membrane forming material and acetic acid glacial (Fisher Scientific) was used as the solvent. All the chemicals were used as received.

3. Methodology

3.1. Membrane fabrication

CA solution was prepared by mixing 23 wt% of CA in 77 wt% of acetic acid at 65 °C for 4 hours at 250 rpm. The solution was cast by using the self-fabricated casting machine at 4.74 m/s [11] once the bubbles was removed. The cast CA solution was placed the flowing coagulant to form the membrane. RO water was used as the coagulant. Three types of coagulant flow were studied, which were laminar, transition and turbulent. The type of flow was determined by calculating the Reynolds number. The membrane was labelled according to the type of flow. CA-Laminar indicates that the CA membrane was immersed in a laminar flow of RO water during the phase inversion process.

3.2. Pure water permeation (PWP)

The flux of membrane was determined by conducting the ultrafiltration experiment at 3 bar using Milipore stirred cell. The effective membrane area was 13.4 cm^2. The experiment was conducted for 3 hours to collect the permeate.

3.3. Tensile test

The mechanical strength of the membrane was determined by using Shimadzu AG-X Universal Tensile Machine, according to the ASTM D882-09 [12].

3.4. Contact angle measurement

3 μL of water was placed on the surface of the membrane using the sessile drop method. Ten contact angle data was collected for each membrane using the Attension Theta Automatic Goniometer.

3.5. Porosity

Gravimetric method was adopted to measure the porosity of the membrane. The porosity was calculated by using the following equation (1):

$$\varepsilon = \frac{(w_a - w_b)/\rho_{water}}{(w_a - w_b)/\rho_{water} + w_b/\rho_{membrane}} \quad (1)$$

where w_a and w_b are the weight of wet membrane and dry membrane respectively. Meanwhile, ρ_{water} refers to the density of water and $\rho_{membrane}$ is the density of membrane

4. Result and discussion

PWP and porosity of the CA membranes are shown in figure 1. When the flow of RO water increased from laminar flow to turbulent flow, PWP of the membrane increased. It was due to higher porosity of the CA-Turbulent membrane compared to CA-Laminar membrane. During phase inversion process, rapid solvent-coagulant exchange between CA solution and RO water resulted a more porous structure in CA-Turbulent [13]. It was due to high solvent concentration gradient between CA solution and RO water [14].

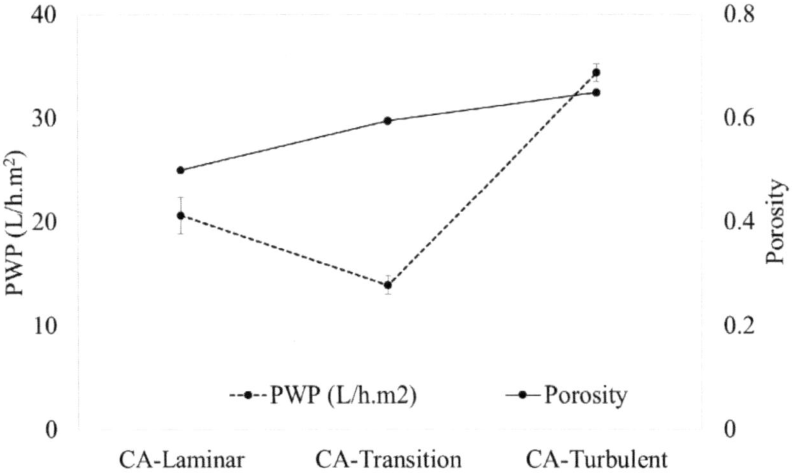

Figure 1. The performance of CA membranes in terms of PWP and porosity.

When the CA solution contacted with RO water, acetic acid would leave the CA solution and mixed with RO due to its high water miscibility. Hence, concentration of acetic acid in the RO water was increased as the contact time increased. This implied that the solvent-coagulant exchange rate would be reduced. However, in turbulent flow, the high concentration gradient between CA solution and RO water was maintained at its optimum condition as fresh RO water was pumped into the system continuously during membrane formation process. Short contact time between CA solution and RO water due to the high velocity of RO water also contributed in maintaining the high concentration gradient condition between these two mediums. Comparatively, during the formation of CA-Laminar membrane, exchange rate was slower and thus porosity was approximately 23% lower compared to CA-turbulent membrane.

Table 1 shows the tensile strength and hydrophilicity of CA membranes. Generally, CA membranes were hydrophilic as the contact angle was less than 90°. Type of flow did not affect the hydrophilicity of membrane as the contact angle was ~73° for all the membranes. In terms of strength, CA-Transition and CA-Turbulent membranes showed ~33MPa where CA-Laminar membrane exhibited the lowest tensile strength, which was ~15MPa. This was due to the presence of large quantity of nodules on the membrane surface, which reduced the surface integrity[14]. Comparatively, strength of the membranes produced in this study was higher than the CA membranes produced by Etemadi et al. [15] and Li et al. [16], which exhibited ~5.2MPa and ~3MPa, due to double layer macrovoids free structure.

Table 1. The physical properties of the CA membranes prepared under different flow conditions.

Membrane ID	**Tensile strength (MPa)**	**Contact angle (°)**	**Description on the morphology of the membrane** [14]
CA-Laminar	15.1886 ± 3.5731	73.0948 ± 3.8080	–Rough, with large quantity of nodules on the

			surface. – Macrovoids free structure
CA-Transition	35.6778 ± 5.8424	72.8586 ± 3.4277	– Rough, with moderate quantity of nodules – Macrovoids free structure
CA-Turbulent	32.2607 ± 4.7077	72.4780 ± 2.7308	– Smooth, with small quantity of nodules – Macrovoids free structure

5. Conclusion
This study showed that the flow of coagulant affect porosity, PWP and tensile strength of CA membranes. Turbulent coagulant produced more porous membrane compared to laminar flow due to rapid solvent-coagulant exchange rate. This explains for the higher PWP in CA-Turbulent membrane compared to CA-Laminar membrane. Results also showed that CA-Turbulent membrane exhibited good strength property due to smooth surface and macrovoids free structure. In conclusion, turbulent coagulant produced smooth porous membrane with good strength property, which is desired for wide range of separation purpose.

Acknowledgements
Financial support from SEGi University (Project number: SEGiIRF/2015-01/FoEBE-15/84) is gratefully acknowledged.

References
[1] 1997 *Environmental Technology Best Practice Programme, A Guide to Cost-Effective Membrane Technologies for Minimising Wastes and Effluents*
[2] J Caro, M Noack, P Kolsch and R Schafer 2000 *Micropor. Mesopor. Mat.* **38** pp 3
[3] W L Chou, D G Yu and M C Yang 2005 *Polymer. Adv. Tech.* **16** pp 600
[4] M Baghbanzadeh, D Rana, C Q Lan and T Matsuura 2016 *Sep. Purif. Rev.* **45** pp 141
[5] M K Chan and A Idris 2012 *J. Ind. Eng. Chem.* **18** pp 2115
[6] S Laohaprapanon, A D Vanderlipe, B T Dome Jr and S J You 2017 *J. Taiwan Inst. Chem. Eng.* **70** pp 15
[7] M G Kochameshki, A Marjani, M Mahmoudian and K Farhadi 2017 *Chem. Eng. J.* **309** pp 206
[8] H T Dang, R M., Narbaitz and T Matsuura 2008 *J. Membrane Sci.* **323** pp 45
[9] Q F Alsalhy, H A Salih, S Simone, M Zablouk, E Drioli and A Figoli 2014 Desalination **345** pp 21
[10] J J Qin and T S Chung 2004 J. Membrane Sci. **229** pp 1
[11] M K Chan and H H Teo 2013 *Malaysia Patent:* PI 2013003337
[12] 2009 *ASTM International, ASTM D* pp 882-09
[13] C A Smolders and A J Reuvers 1992 *J. Membrane Sci.* **73** pp 259
[14] M K Chan, P Raoo, K E Yap and J M Y Lim 2016 *Mater. Manuf. Process.* **31** pp1941
[15] H Etemadi, R Yegani and V Babaeipour 2016 *Diamond Relat. Mater.* **69** pp 166
[16] G Li, J Wang, D Hou and Y Bai, H Liu 2016 *J. Environ. Sci. (China)* **45** pp 7

Facial synthesis of nanostructured $ZnCo_2O_4$ on carbon cloth for supercapacitor application

Swati J Patil, Jungsung Park and Dong-Weon Lee*

School of Mechanical Systems Engineering, Chonnam National University, 77 Yongbong-ro, Buk-gu, Gwangju 500757, Republic of Korea.

*E-mail: mems@jun.ac.kr

Abstract. In this work, we have synthesized the $ZnCo_2O_4$ electrode by a facial one-step hydrothermal method on a carbon cloth for the supercapacitor application. The structural and phase purity of the prepared electrode material was confirmed by X-ray diffraction (XRD) technique. The surface morphology and elemental stoichiometry were studied using field emission scanning electron microscopy (FE-SEM). The FE-SEM micrograph illustrates that the $ZnCo_2O_4$ material is composed of microstrips with a ~0.5 μm width and length in micron uniformly covered the carbon cloth surface. The $ZnCo_2O_4$ electrode material further investigated for electrochemical analyses. The cyclic voltammetry results showed that the $ZnCo_2O_4$ microstrips electrode exhibited the highest specific capacitance of 1084 F/g at 2 mV/s scan rate. Remarkably, a maximum energy density of 12.5 Wh/kg was attained at a current density of 2 mA/cm^2 with the power density of 3.6 kW/kg for the $ZnCo_2O_4$ microstrips electrode. Furthermore, the 96.2 % capacitive retention is obtained at a higher scan rate of 100 mV/s after 1000 CV cycles, indicating excellent cycling stability of the $ZnCo_2O_4$ microstrips electrode. The frequency-dependent rate capability and an ideal capacitive behaviour of the $ZnCo_2O_4$ microstrips electrode were analyzed using impedance analyses; a representing the ion diffusion structure of the material. These results show that the $ZnCo_2O_4$ microstrips electrode could be a promising material for supercapacitor application.

1. Introduction

Supercapacitors with extraordinary features including high power density, high energy density, and long term cyclic lifespan are attracted and it can use for advanced green energy storage applications such as electric vehicles, portable electronic devices, etc [1]. Supercapacitors can divide into mainly two types: electrical double layer capacitors (EDLCs) and pseudocapacitors that based on the charge storage mechanism. EDLCs type of capacitor, the electrostatic store charges *via* reversible adsorption of electrolytic ions at the electrode and electrolyte interface to form electric double layer capacitors. While pseudo capacitor store charge *via* reversible faradic redox reaction [2]. Mostly, carbon-based materials are assisted as electrode materials of EDLCs type. Typically metal oxides/hydroxides and polymers served as pseudo-capacitive materials. Recently, Spinel oxide type of transition metal oxides is much attracted for electrochemical application with differing nanostructure architecture. Compared with the monometallic oxides, Spinel oxides exhibiting synergetic effect from two metal ions and effectively achieves the good electrochemical rate performance [3]. Transition metal substituted Spinel structures of cobaltite oxide MCo_2O_4 (M = Mn, Ni, Cu, and Zn) are most promising for current charge storage applications [4-7]. These types of mixed transition metal oxides offer higher electronic conductivity as compared to single metal oxides. Among the Spinel type transition metal oxide, zinc cobaltite ($ZnCo_2O_4$) electrode material is one of the attractive Spinel structure offer richer redox activities, good electrical conductivity, low cost and environmentally friendly [8]. Moreover, the cobalt and zinc two metal cations provide the higher capacity and an effective electron transport pathway, respectively in the $ZnCo_2O_4$ material [9].

In this work, the one-step hydrothermal method was employed for the preparation of nanostructured the $ZnCo_2O_4$ microstrips on a carbon cloth. The structural, morphological and chemical composition characteristics were systematically examined using XRD and FE-SEM techniques, respectively. Further, electrochemical measurements were investigated using cyclic voltammetry (CV) and charge/discharge (CD) study. Electrochemical impedance spectroscopy (EIS) was used to study the interface behavior of electrode. The $ZnCo_2O_4$ microstrips electrode exhibited excellent electrochemical performance with superior cycling stability of 96.2 % up to 1000 CV cycles.

2. Experimental studies

All the chemicals were used without further any purification. The $ZnCo_2O_4$ material was prepared by facial hydrothermal method on carbon cloth. For a preparation of $ZnCo_2O_4$, 4 mmol ammonium fluoride, 1 mmol Zinc acetate, and 2 mmol cobalt nitrate were dissolved into the 45 ml deionized (DI) water. After stirring for 45 min, the resulting solution was moved to a Teflon-lined stainless steel autoclave followed by maintaining at 150 °C for 12 h. Then the precipitates were collected, cleaned with DI water and ethanol alternatively and dried in a vacuum oven at 70 °C for overnight. Finally, the resultant product was calcinated at 400 °C for 2 h to remove the organic residue in the furnace and then used for characterizations.

2.1. Material characterization techniques

The structural and phase purity of the prepared material was examined by powder X-ray diffraction (XRD) with Cu Kα radiation. The surface morphological and elemental compositions of the prepared material were investigated by field emission scanning electron microscope (FE-SEM). The electrochemical properties of the prepared electrode were carried out using cyclic voltammetry, galvanostatic charge/discharge and impedance analysis by the compatible IVIUM technologies electrochemical workstation. The electrochemical measurements were investigated in three-electrode cell configuration (saturated calomel electrode as reference and platinum as a counter electrode) in 1 M KOH electrolyte with 0 to 0.5 V/SCE scanning sweep potential ranges.

The areal and specific capacitances of the $ZnCo_2O_4$ microstrips electrode was calculated using equation (1). Where, Ca and Cs are the areal (F/cm^2) and specific capacitance (F/g), respectively of electrode material. I, v and (V2−V1) are the current response (mA), scan rate (mV/s) and the working potential window (V/SCE) of in the unit area (1 cm^2). m is the areal mass of electrode material deposited on carbon cloth. Further, the discharging capacitance (Ccd, F/g), specific energy (E, Wh/kg), and specific power density (P, W/kg) were calculated using equation (2), as follows [10];

$$C_a = \frac{\int_{V_1}^{V_2} I(V)dV}{v \times (V_2 - V_1)} \quad \text{and} \quad C_s = \frac{C_a}{m} \tag{1}$$

$$C_{cd} = \frac{I \times Td}{(V_2 - V_1) \times m}, \quad E = \frac{0.5 \times C_{cd} \times (V_2^2 - V_1^2)}{3.6} \quad \text{and} \quad P = \frac{E \times 3600}{Td} \tag{2}$$

3. Results and discussion

3.1. Structural and morphological studies

Figure. 1(a) displays the XRD pattern of $ZnCo_2O_4$ material prepared on carbon cloth. The major diffraction peaks observed at 31.47°, 36.96°, 45.06°, 59.59° and 65.18° are indexed as (220), (311), (400), (511) and (440) planes of the cubic $ZnCo_2O_4$ structure. The obtained XRD peaks best matches with standard diffraction pattern of $ZnCo_2O_4$ (JCPDS card: 00-001-1149) as shown in figure 1(a) bottom. Moreover, the broad peak was found at the peak position of 25° related to activated carbon [11]. From XRD study it is clear that $ZnCo_2O_4$ material exhibits a polycrystalline nature. To identify the elemental composition, Energy-dispersive X-ray spectroscopy (EDX) was used. Figure. 1(b) shows the elemental spectrum and inset table displays elemental compositions of $ZnCo_2O_4$ material. From the EDX analysis, it demonstrates the existence of Zn, Co and O elements with the elemental composition ratio of 1:2:4, respectively. Further surface microstructural images were analyzed using FE-SEM technique. Figure 1 (c, d) show the typical FE-SEM images of $ZnCo_2O_4$ material on carbon cloth at ×10 k and ×50 k magnifications. Figure 1 (d) clearly reveling that microstrip-like surface morphology with strip width nearly 0.5 μm covered the carbon cloth surface uniformly. The microstrips have a porous surface with nanocrystalline grains. Such porous morphological surface important is for ion adsorption in the redox reaction to store more charge（5）.

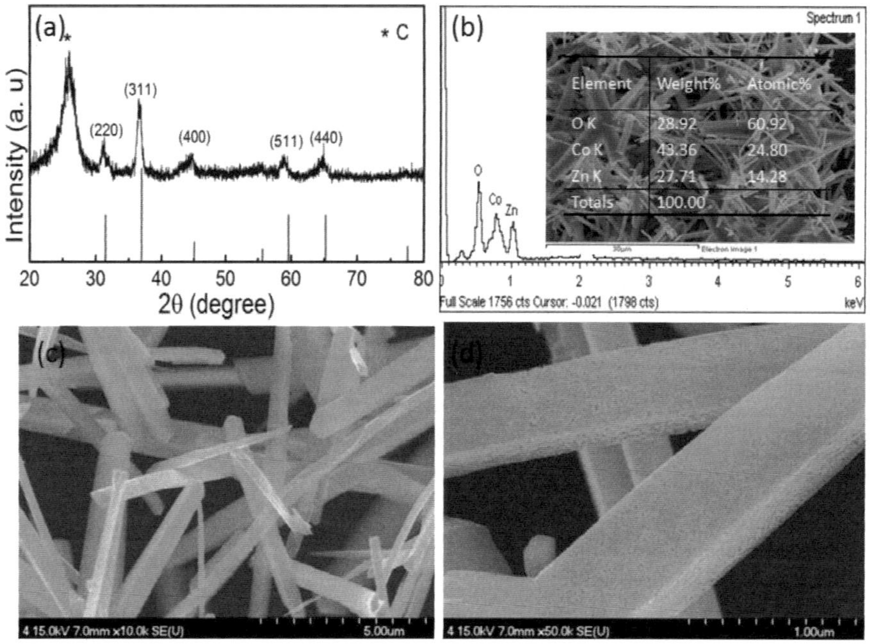

Figure 1. (a) XRD pattern of the $ZnCo_2O_4$ electrode grown on carbon cloth and bottom part shows stick pattern corresponding to JCPDS card (00-001-1149). **(b)** EDX spectrum with inset shows the elemental composition of the $ZnCo_2O_4$ electrode. **(c)** and **(d)** FESEM images of the $ZnCo_2O_4$ electrode at × 10 k and ×50 k magnifications, respectively.

3.2. Electrochemical studies

The cyclic voltammetry measurements were carried out at various scan rates (2, 5, 10, 20, 50, 70, 100, 150 to 200 mV/s) and results as shown in figure 2. From the CV curves, it can be noticed that the shape of CV curves remain unchanged at different scan rates. As scan rate increase, the peak current increases. The areal and specific capacitance of the $ZnCo_2O_4$ electrode area calculated from equation (1). Figure 2(a) and (b) shows that areal and specific capacitance decreased with increasing scan rate. The highest areal and specific capacitances at 2 mV/s scan rate are found to be 0.13 F/cm^2 and 1084 F/g, respectively. Even at higher scan rates of 100 and 200 mV/s, the specific capacitance is of 225 and 173 F/g, respectively indicating a good rate capability of the $ZnCo_2O_4$ electrode in the aqueous electrolyte. The specific capacitance decreased from 751 to 225 F/g at 10 to 100 mV/s scan rate may be due to the effective inner surface adsorption of ions was reduced at higher scan rate accordingly. At higher scan rates, the decrease of the specific capacitance is attributed to the diffusion ion charge transports lagging behind the electronic mobility [12]. Venkatachalam et al [13] reported the specific capacitance of 845.7 F/g for the double hydroxide treated the $ZnCo_2O_4$. The electrochemical cyclic stability rate performance of the $ZnCo_2O_4$ electrode carried out at a 100 mV/s scan rate over 1000 CV cycles. As seen in figure. 2(d), area under the curve little decreased showing the good cyclic stability of the $ZnCo_2O_4$ electrode in KOH electrode. It is seen that capacitive retention of the $ZnCo_2O_4$ electrode as a function of cycle number. The capacitive retention of 96.2 % retained after 1000 cycles for the $ZnCo_2O_4$ electrode that is better than coral-$ZnCo_2O_4$ nanowires with 85% capacitance retention after 2000 cycles [8]. The capacitive retention of the $ZnCo_2O_4$ microstrips electrode decreased after hundred of CV cycles possibly due to the initial CV cycles that induce a lower rate of ion adsorption.

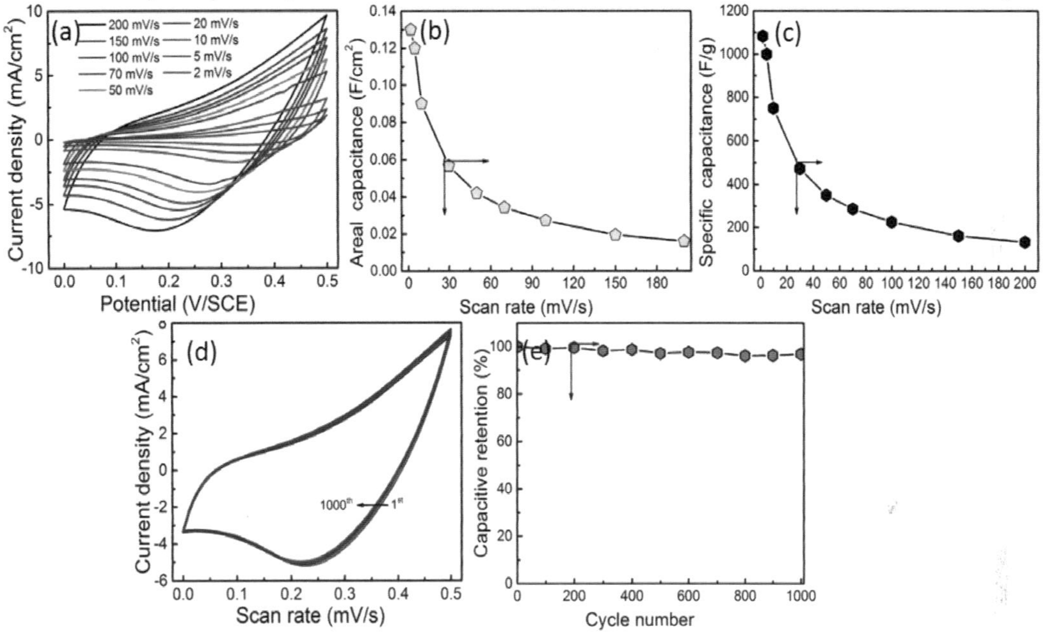

Figure 2. Electrochemical studies of the $ZnCo_2O_4$ microstrips electrode: (**a**) CV curves at different scan rates (from 2 to 200 mV/s) and corresponding plots show the areal **(b)** and specific **(c)** capacitances. **(d)** The electrochemical cyclic stability CV curves of the $ZnCo_2O_4$ microstrips electrode with various cycle numbers, and **(e)** the capacitive retention against a number of cycles.

Figure 3. (a). shows the charge/discharge profiles of the $ZnCo_2O_4$ microstrips electrode measured at 2, 5, 7 and 10 mA/cm^2 current densities. The charge/discharge time reduced with an increase in the current density. From the charge/discharge curves, the semi-symmetric nature of the $ZnCo_2O_4$ electrode reveals that typical pseudo capacitive behavior. The discharge capacitance under different current densities was calculated from equation (2). The discharge capacitance as a function of current density is plotted in figure 3(b). The maximum discharge capacitance is 423 F/g and the corresponding areal capacitance is 0.05 F/cm^2. The discharge capacitance of the $ZnCo_2O_4$ microstrips electrode decreased from 423 at 2 mA/cm^2 to 112 F/g at 10 mA/cm^2 current densities. It means that the capacitance of the $ZnCo_2O_4$ microstrips electrode as a function of current density. Further, using equation (2) the energy and power density were calculated. The Ragone plot as seen in figure 3(c), it suggests that the specific power density increases with decreasing energy density. The calculated maximum energy density of the $ZnCo_2O_4$ microstrips electrode is about 12.5 Wh/kg with the delivered power density of 3.6 kW/kg. The obtained values of energy and the power density of the $ZnCo_2O_4$ microstrips electrode are higher than $ZnFe_2O_4$/NRG composite [14]. It means that, the $ZnCo_2O_4$ store 6.7 Wh/kg energy density with delivering a power of 3000 W/kg.

The electrochemical surface interface behavior and ion diffusion rate of the $ZnCo_2O_4$ electrode were evaluated using impedance analysis. Figure 3(d and e) displays impedance measurement results of the $ZnCo_2O_4$ electrode (frequency range: 1 MHz to 0.1 Hz). The Nyquist plot, the real impedance against imaginary impedance of the $ZnCo_2O_4$ microstrips electrode as presented in figure 3(d). Nyquist plot divided into two frequency regions. In the high-frequency region, a semicircle and a straight line occurred in the low-frequency region. The occurrence of a vertical line along with the imaginary axis indicates the ideal capacitive behaviour of the $ZnCo_2O_4$ electrode and the electrode material has a lower ion diffusion resistance [15]. The internal solution resistance (Rs) and charge-transfer resistance (Rct) values of the $ZnCo_2O_4$ electrode are found to be 1.9 Ω and 3.5Ω, respectively indicating that good charge transfer performance of electrode material. The frequency dependent phase angle plot as presented in figure. 3(e). It shows that the phase angle approaches -75° at 79 Hz frequency. Furthermore, the Bode plot magnitude plot (figure. 3(e)) suggests that in the lower frequency region the $ZnCo_2O_4$ microstrips electrode shows an ideal capacitive behavior.

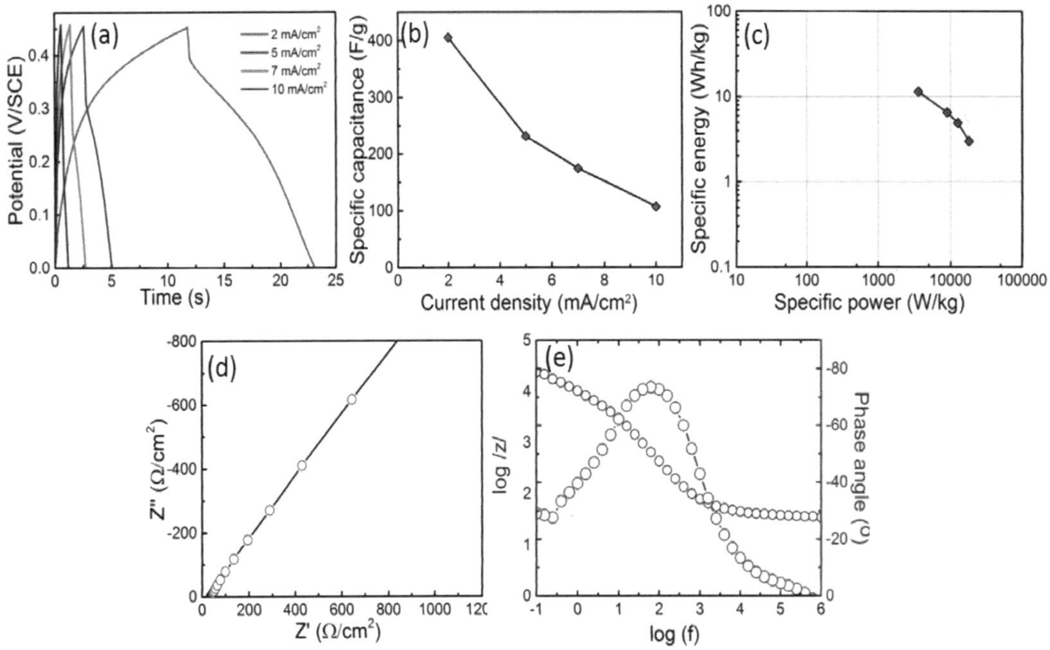

Figure 3(a-b). Charge/discharge profiles with various current and corresponding the specific capacitance of the $ZnCo_2O_4$ microstrips electrode. **(c)** Ragone plot of the $ZnCo_2O_4$ microstrips electrode at various current densities. Impedance analysis: **(d)** Nyquist plot and **(e)** Bode magnitude/ phase plots of the $ZnCo_2O_4$ microstrips electrode.

4. Conclusion

In conclusion, the $ZnCo_2O_4$ have been prepared *via* a hydrothermal method followed by calcination at 400 °C. Further, structural and morphological studies systematically studied. The electrochemical study proved that $ZnCo_2O_4$ microstrips electrode has good rate capability with 1084 F/g specific capacitance and showed the superior capability of 96.2% capacitive retention in 1 M KOH aqueous electrolyte. These results indicate that the $ZnCo_2O_4$ microstrip architecture desirable for long term cyclic stability rate performance in the electrochemical application.

Acknowledgment

This work was supported by the National Research Foundation of Korea (NRF) grant funded by the Korea government (MSIP) (No. 2015R1A4A1041746).

References

[1] Miller J R and Simon P 2008 *Science* **321** pp 651–52
[2] Yu Z, Tetard L, Zhai L and Thomas J 2015 *J Energy Environ. Sci.* **8** pp 70-730
[3] Pendashteh A, Palma J, Anderson M and Marcill R 2015 *J. Mater. Chem. A* **3** pp 16849-59
[4] Rajeshkhann G, Umeshbabu E, Justin P and Rao R G 2017 *J. Alloys Compd.* **696** pp 947–55
[5] Patil S J, Kim J H and Lee D W 2017 *Chem. Eng. J.* **322** pp 498–509
[6] Mohamed S G, Hung T F, Chen C J, Chen C K, Hu S F and Liu R S 2014 *RSC Adv.* **4** pp 17230–35
[7] Dubal D P, Gomez-Romero P, Sankapal B R and Holze R 2015 *Nano Energy* **11** pp 377–99
[8] John A R, Min B-K, Kim J-H, Kang S-H, Kim H and Ahn K-S 2017 *J. Electroanal. Chem.* **785** pp 48–57
[9] Xu L, Zhao Y, Lian J, Xu Y, Bao J, Qiu J, Xu L, Xu H, Hua M and Li H 2017 *Energy* **123** pp 296–304
[10] Patil S J, Kim J H and Lee D W 2017 *J. Power Sources* **342** pp 652-65
[11] Guo D, Ren W, Chen Z, Mao M, Li Q and Wang T 2015 *RSC Adv.* **5** pp 10681-687
[12] Tomboc G M, Jadhav H S and Kim H 2017 *Chem. Eng. J.* **308** pp 202–13
[13] Venkatachalam V, Alsalme A, Alswieleh A and Jayavel R 2017 *Chem. Eng. J.* **321** pp 474–83
[14] Li L, Bi H, Gai S, He F, Gao P, Dai Y, Zhang X, Yang D, Zhang M and Yang P 2017 *Sci. Rep.* **7** pp 43116
[15] Wang K P and Teng H 2007 *J. Electrochem. Soc.* **154** pp A993–A98

IOP Conf. Series: Materials Science and Engineering **282** (2017) 012005 doi:10.1088/1757-899X/282/1/012005

Dual lead-crowning for helical gears with anti-twist tooth flanks on the internal gear honing machine

Van-Quyet Tran and Yu-Ren Wu*

Department of Mechanical Engineering, National Central University, 300, Jhongda Rd., Jungli City, Taoyuan 320, Taiwan, R.O.C.

*E-mail: yurenwu@ncu.edu.tw

Abstract. For some specific purposes, a helical gear with wide face-width is applied for meshing with two other gears simultaneously, such as the idle pinions in the vehicle differential. However, due to the fact of gear deformation, the tooth edge contact and stress concentration might occur. Single lead-crowning is no more suitable for such a case to get the appropriate position of contact pattern and improve the load distribution on tooth surfaces. Therefore, a novel *Email: method is proposed in this paper to achieve the wide-face-width helical gears with the dual lead-crowned and the anti-twisted tooth surfaces by controlling the swivel angle and the rotation angle of the honing wheel respectively on an internal gear honing machine. Numerical examples are practiced to illustrate and verified the merits of the proposed method.

1. Introduction

In some specific applications, the helical gears with wide face-width are applied for meshing with another two gears simultaneously, such as those idle pinions in the helical gear-type vehicle limited slip differential (EATON Detroit Truetrac®). For such a case, one half of the tooth flank of a pinion meshes with one side gear and another half meshes with one other pinion for distributing the transmission power. However, the gear tooth deformation and deflection might occur due to the large torque, leading to the possibilities of appearance of the tooth edge contact and early gear damage. Based on evidence that the helical gears with crowned and twist-free tooth flanks can enhance the load capacity, reduce the risk for high concentrated contact pressure and extend life-span, how to achieve the tooth modification on two half tooth flanks for this case is really worthy to study further.

In the past, several studies about the gear honing have been presented. Dugas [1] and [2] presented the research about rotary gear honing such as working process of honing machine and honing tool. Wright and Schriefer [3] presented some discussions about the gear honing process as well as advantages of internal gear honing. Klocke [4] introduced axes motions and combinations on the internal gear honing machine for removing material of tooth flank on the work gear surface. Mehta and Rathi [5] mentioned about the internal gear honing is the most efficient process of gear finishing. Besides, Amini et al. [6] conducted an experimental study of gear abrasive wear for reducing noise.
Moreover, some published studies mentioned about single lead-crowning for helical gear. Tran et al. [7]–[8] proposed a crowning method by controlling an additional rotation angle of the work gear during the hobbing process. Further, Wu and Tran [9]–[10] proposed a method by controlling the

crossed angle between axes of the honing cutter and the work gear to achieve the crowned and twist-free helical tooth flanks using a variable pressure angle (VPA) honing wheel.

Referring to the aforementioned studies and the encountered problem, a novel method to obtain a helical gear with dual lead-crowned tooth flanks using the internal gear honing process is proposed in this paper. Two additional motions are given for achieving the dual lead-crowning and the twist-free tooth surfaces by setting the swivel angle of honing wheel a linear function of axial feed of honing wheel and by adding the rotation angle of work gear a product of the rotation angle of honing wheel and the axial feed of honing wheel with a coefficient, respectively. Based on the proposed method, two numerical examples are carried out to prove the practicability and validity of this method.

2. Mathematical model of generation of standard honing wheel

As shown in figure 1, the position vector and the unit normal vector on right surface of the rack can be expressed in the coordinate system $S_r\,(x_r, y_r, z_r)$ as follows:

$$\mathbf{r}_r(u,v)=\left[x_r, y_r, z_r, 1\right]^T=\left[u\cos\alpha_{pa}, w-u\sin\alpha_{pa}, v, 1\right]^T \tag{1}$$

$$\mathbf{n}_r(u,v)=\left[x_{nr}, y_{nr}, z_{nr}\right]^T=\left[\sin\alpha_{pa}, -\cos\alpha_{pa}, 0\right]^T \tag{2}$$

where u and v are the surface parameters of rack, α_{pa} is the pitch pressure angle of standard rack and w is a quarter of tooth pitch and can be calculated by $w=\pi m_n/4$, wherein m_n is the normal module.

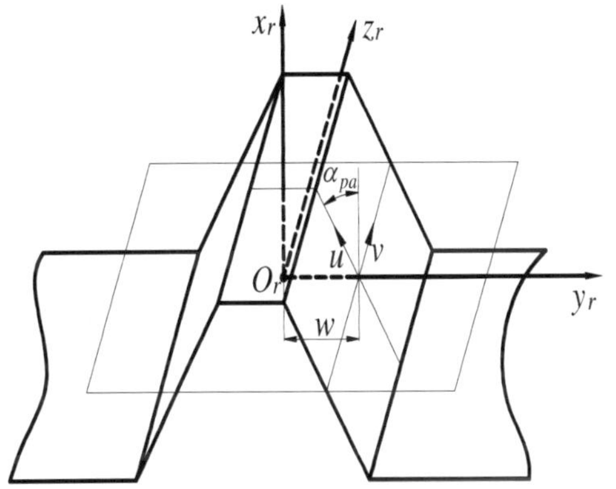

Figure 1. Geometric schematics of the rack.

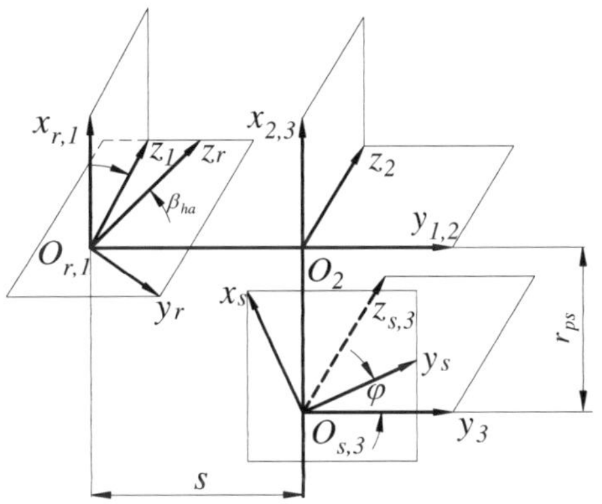

Figure 2. Relative motion coordinate systems between rack and helical gear.

As shown in figure 2, the helical gear can be generated by rack according to the relative motion relationship. Where the coordinate systems $S_r\,(x_r, y_r, z_r)$, $S_s\,(x_s, y_s, z_s)$ and $S_3\,(x_3, y_3, z_3)$ are rigidly attached to the rack, helical gear and frame, respectively. Then, the equations of locus and unit normal vector of rack in S_s can be obtained by transforming from S_r to S_s as follows:

$$\mathbf{r}_s(u,v,\varphi)=\mathbf{M}_{sr}(\varphi)\cdot\mathbf{r}_r(u,v) \tag{3}$$

$$\mathbf{n}_s(u,v,\varphi)=\mathbf{L}_{sr}(\varphi)\cdot\mathbf{n}_r(u,v) \tag{4}$$

where $\mathbf{M}_{sr}(\varphi)$ is the transformation matrix from S_r to S_s, $\mathbf{L}_{sr}(\varphi)$ is the upper-left 3×3 submatrix of $\mathbf{M}_{sr}(\varphi)$. Symbols φ and r_{ps} are the rotation angle and operating pitch radius of the generated helical gear, respectively. And s is the displacement of rack, it can be calculated by $s=r_{ps}\varphi$.

According to the theory of gearing (Litvin **[11]**), the meshing condition between the right rack surface and the right helical gear surface can be established by:

$$f_1(u,v,\varphi)=\mathbf{n}_s\cdot\frac{\partial\mathbf{r}_s(u,v,\varphi)}{\partial\varphi}=0 \tag{5}$$

As shown in figure 3, the honing wheel can be generated by the helical gear using the homogenous coordinate transformation from S_s to S_h (the moving coordinate system of honing wheel). The equations of locus and unit normal vector of rack can be represented in S_h as follows:

$$\mathbf{r}_h(u,v,\varphi,\theta_h)=\mathbf{M}_{hs}(\theta_h)\cdot\mathbf{r}_s(u,v,\varphi) \tag{6}$$

$$\mathbf{n}_h(u,v,\varphi,\theta_h)=\mathbf{L}_{hs}(\theta_h)\cdot\mathbf{n}_s(u,v,\varphi) \tag{7}$$

where $\mathbf{M}_{hs}(\theta_h)$ is the transformation matrix from S_s to S_h, $\mathbf{L}_{hs}(\theta_h)$ is the upper-left 3×3 submatrix of $\mathbf{M}_{hs}(\theta_h)$. Symbols θ_s and θ_h are the rotation angle of helical gear and honing wheel, respectively. Besides, θ_h can be calculated by $\theta_h = N_s\varphi/N_h$, wherein N_s and N_h are the tooth number of helical gear and honing wheel, respectively. And r_{ps} and r_{ph} are operating pitch radii of generated helical gear and honing wheel, respectively. The symbol E_o is the centre distance between helical gear and honing wheel.

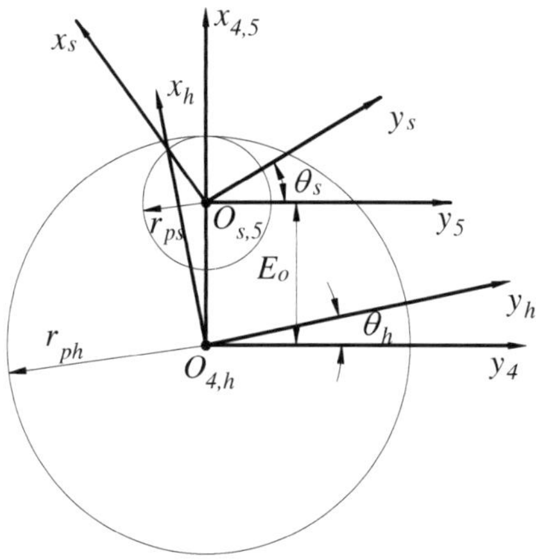

Figure 3. Relative motion coordinate systems between helical gear and honing wheel.

The meshing condition between the right helical gear surface and right honing wheel surface can be obtained by:

$$f_2(u,v,\varphi,\theta_h)=\mathbf{n}_h\cdot\frac{\partial\mathbf{r}_h(u,v,\varphi,\theta_h)}{\partial\theta_h}=0 \tag{8}$$

Finally, the equation of honing wheel surface can be obtained by solving Eq.s (5), (6) and (8) simultaneously.

3. Dual lead-crowning of helical gear on internal gear honing machine

3.1 Mathematical model

As shown in figure 4, the 3-D schematic of an internal gear honing machine (Fässler HMX-400) shows eight moving axes for gear honing, wherein T_1, T_2 and T_3 are three translational motion axes in the radial, tangential and axial feed axes respectively, R_1 is the swivel axis of honing-stone head for lead crowning, R_2 is the swivel axis of the honing stone for adjusting the axis-crossing angle, R_3 is the swivel axis of headstock turntable, R_4 is the spindle axis of workpiece and R_5 is the spindle axis of honing wheel. This machine is then used in this paper to practice the dual lead-crowning on helical gears.

The relative motion coordinate systems between the honing wheel and the workpiece are established in figure 5, the coordinate systems S_c (x_c, y_c, z_c) and S_w (x_w, y_w, z_w) are rigidly attached to the honing wheel and work gear, respectively, the coordinate system S_8 (x_8, y_8, z_8) is rigidly attached to the frame of honing machine, while $S_6(x_6, y_6, z_6)$, $S_7(x_7, y_7, z_7)$, $S_9(x_9, y_9, z_9)$ and $S_d(x_d, y_d, z_d)$ are auxiliary coordinate systems for the simplification of coordinate transformation.

For lead-crowning of a work gear, a linear function of axial motion of the work gear is given to the swivel angle of honing wheel around the R_2 axis, ϕ_w, as the following equation:

$$\phi_w = kt \tag{9}$$

where k is an adjusting coefficient of the swivel angle, and t is honing wheel's feed along T_2 axis.

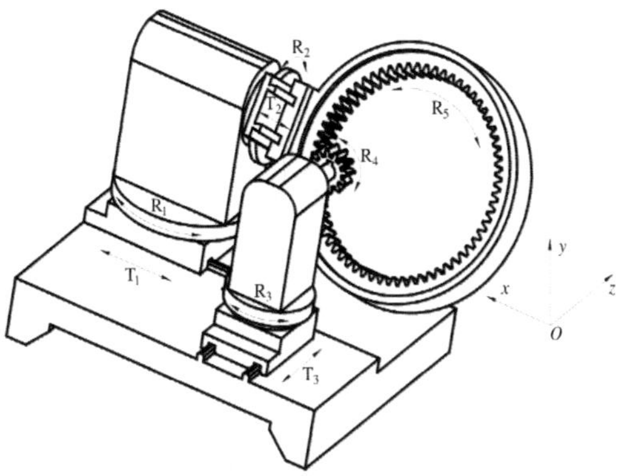

Figure 4. 3-D schematic of a CNC internal gear honing machine.

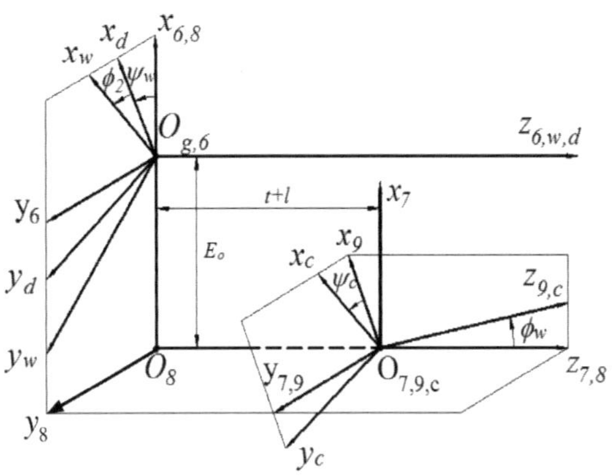

Figure 5. Coordinate systems for Fässler HMX-400 internal gear honing machine.

By applying the homogeneous coordinate transformation from S_c to S_w, the equations of locus and unit normal vector of the honing wheel can be represented in the coordinate system S_w as follows:

$$\mathbf{r}_w(u,v,\varphi,\theta_h,\psi_c,t) = \mathbf{M}_{wc}(\psi_c,t)\cdot\mathbf{r}_h(u,v,\varphi,\theta_h) \tag{10}$$

$$\mathbf{n}_w(u,v,\varphi,\theta_h,\psi_c,t) = \mathbf{L}_{wc}(\psi_c,t)\cdot\mathbf{n}_h(u,v,\varphi,\theta_h) \tag{11}$$

where $\mathbf{M}_{wc}$ is the transformation matrix from S_c to S_w, and $\mathbf{L}_{wc}$ is the upper-left 3x3 submatrix of $\mathbf{M}_{wc}$. All elements in $\mathbf{M}_{wc}$ are listed as follows:

$$\mathbf{M}_{wc} = \begin{bmatrix} a_{11} & a_{12} & a_{13} & a_{14} \\ a_{21} & a_{22} & a_{23} & a_{24} \\ a_{31} & a_{32} & a_{33} & a_{34} \\ 0 & 0 & 0 & 1 \end{bmatrix} \tag{12}$$

wherein, $a_{11} = \cos\phi_w \cos\psi_c \cos(\phi_2 - \psi_w) + \sin\psi_c \sin(\phi_2 - \psi_w)$,
$a_{12} = \cos\psi_c \sin(\phi_2 - \psi_w) + \cos\phi_w \cos(\phi_2 - \psi_w)$, $a_{13} = \sin\phi_w \cos(\phi_2 - \psi_w)$, $a_{14} = -E_o \cos(\phi_2 - \psi_w)$,
$a_{21} = \cos\phi_w \cos\psi_c \sin(\phi_2 - \psi_w) + \sin\psi_c \cos(\phi_2 - \psi_w)$, $a_{22} = \cos\psi_c \cos(\phi_2 - \psi_w) - \sin\phi_w \sin(\phi_2 - \psi_w)$,
$a_{23} = \sin\phi_w \sin(\phi_2 - \psi_w)$, $a_{24} = -E_o \sin(\phi_2 - \psi_w)$, $a_{31} = -\sin\phi_w \cos\psi_c$, $a_{32} = \sin\phi_w \sin\psi_c$, $a_{33} = \cos\phi_w$ and $a_{34} = t + \ell$.

In the above equations, the symbols ψ_w and ψ_c are the rotation angles of the work gear and the honing wheel, respectively, and ϕ_2 is the rotation angle of work gear and ℓ is the distance between two central cross-sections where the dual lead-crowning starts (see in figure 6).

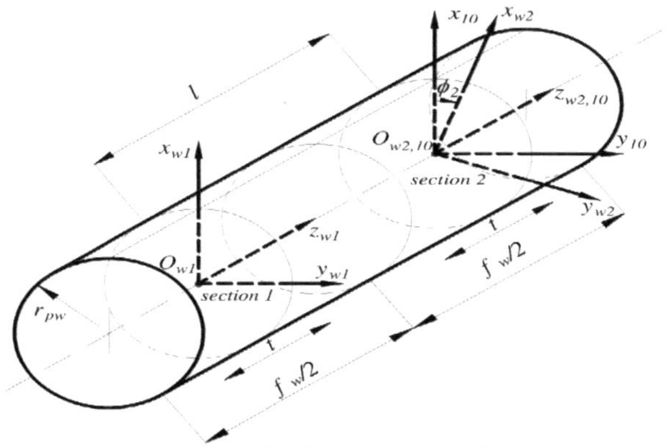

Figure 6. Dual lead-crowning on wide-face-width work gear during honing process.

As shown in figure 6, the dual lead-crowning process can be performed through the following strategy: the tooth flank of work gear is considered to be divided into two parts with the same length. For honing the first half of tooth flank, the honing wheel moves along the axis of work gear and meshes with the work gear at the position of section 1, where $\phi_2 = 0$ and $a_{34} = t$ ($\ell = 0$) are given; and then for the another half of tooth flank, the honing wheel moves axially and meshes with work gear at the position of section 2, where $\phi_2 = f_w / (2r_{pw} \cos\beta_{ha})$ and $a_{34} = t + f_w/2$ ($\ell = f_w / 2$) are defined, wherein f_w is the face width of tooth gear, r_{pw} is the operating pitch radius of work gear and β_{ha} is the helical angle of honing wheel.

The meshing conditions between the right honing surface and right work gear surface can be obtained by the following two equations:

$$f_3(u,v,\varphi,\theta_h,\psi_c,t) = \mathbf{n}_w \cdot \frac{\partial \mathbf{r}_w(u,v,\varphi,\theta_h,\psi_c,t)}{\partial \psi_c} = 0 \tag{13}$$

$$f_4(u,v,\varphi,\theta_h,\psi_c,t) = \mathbf{n}_w \cdot \frac{\partial \mathbf{r}_g(u,v,\varphi,\theta_h,\psi_c,t)}{\partial t} = 0 \tag{14}$$

The rotation angle of work gear, $\psi_w(\psi_c,t)$ can be calculated according to the following equation:

$$\psi_w = \frac{N_c}{N_w}\psi_c + \tan\beta_{ha}\left(\frac{kr_{ph}-1}{r_{pw}}\right)t + b\psi_c t \tag{15}$$

where N_c and N_w denote tooth numbers of the honing wheel and the work gear respectively, ψ_c and ψ_w are the rotation angles of honing wheel and work gear respectively, and b is the correction coefficient for generating the twist-free tooth surface.

3.2 Normal deviation of the work gear tooth surface

The normal deviation of work gear tooth flank can be obtained by comparing the machined gear tooth flank with the standard gear tooth flank. The normal deviations at each grid points can be computed by the following equation:

$$\delta_{ij} = (\mathbf{r}_{w.ij} - \mathbf{r}_{s,ij}) \cdot \mathbf{n}_{s,ij}, \; i = 1\sim13, j = 1\sim13 \tag{16}$$

To obtain dual lead-crowned and anti-twist tooth flanks for the work gear, a flowchart for determining coefficients k and b is given as figure 7 shows.

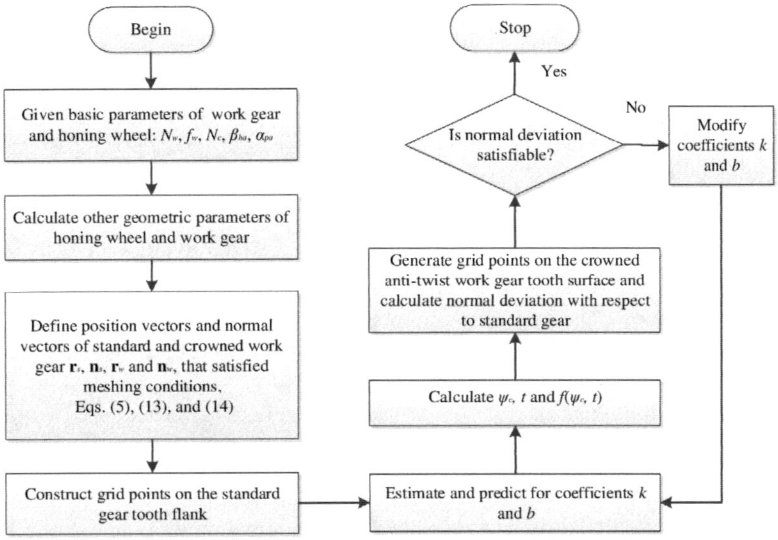

Figure 7. Flowchart for determination of coefficients *k* and *b*.

4. Numerical examples

4.1. Dual lead-crowing for a helical gear with anti-twisted tooth surfaces by gear honing process

To demonstrate the proposed method, two cases with different coefficients $b = 0$ (Case A) and coefficient $b = -2.75\times10^{-3}$ (Case B) and the same coefficient $k = -3.50\times10^{-3}$ are carried out to show two different dual lead-crowned gear tooth surfaces, which are with and without anti-twisted tooth flank corrections, in forms of tooth surface topographies. The related parameters for this example are given in table 1.

Table 1. Related parameters of work gear and honing wheel.

Work gear	**Value [Unit]**
Number of teeth (N_g)	12
Normal module (m_n)	2.0 [mm]
Normal pressure angle (α_{pa})	30.0 [degree]
Pitch helix angle (β_{ha})	20.0 [degree]
Face width (f_w)	40.0 [mm]
Honing wheel	
Number of teeth (N_h)	120
Normal module (m_n)	2.0 [mm]

Moreover, two indices are proposed for evaluating the twist degree of machined gear tooth surfaces, wherein one is the maximum tooth flank twist, M_{tf}, which is defined by the largest crowning amount subtract the smallest crowning amount; and another one is the crowning evenness ratio, C_{er}, which is defined by the smallest crowning amount divided by the largest crowning amount. For an anti-twisted tooth surface, M_{tf} is as smaller as better and C_{er} is as larger as better.

To validate the proposed dual lead-crowning method, the topographies of work gear tooth surfaces without applying an anti-twist tooth surface correction (Case A) and with applying an anti-twist tooth surface correction (Case B) are practiced. As the comparison shown in figure 8, the index M_{tf} for the first part (upper parts) lead-crowned tooth surface is reduced from 16.08 μm (Case A) to 4.59 μm (Case B) and C_{er} is improved from 0.23 (Case A) to 0.71 (Case B). Besides, for the second part (lower parts) tooth surface, M_{tf} is reduced from 15.86 μm (Case A) to 6.39 μm (Case B) and C_{er} is improved from 0.21 (Case A) to 0.59 (Case B). From the results of these two cases, the proposed method is proved that the anti-twisted gear tooth surface can be attained in the dual lead-crowning process.

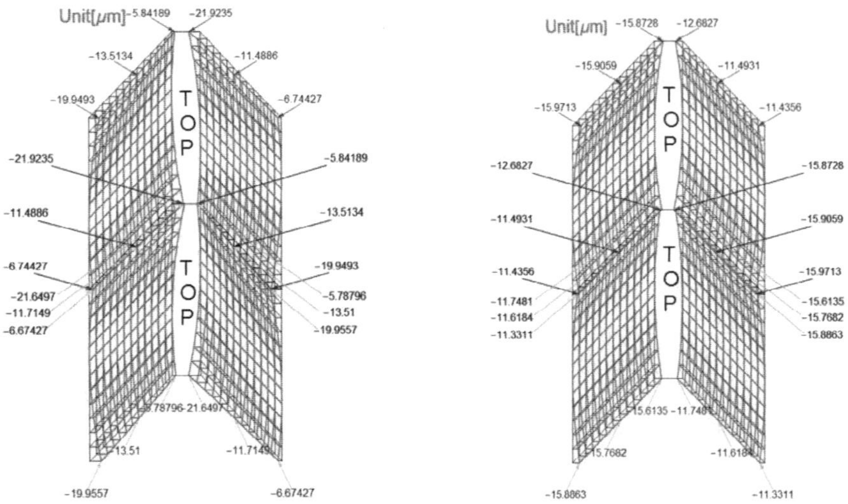

(a) (b)

Figure 8. Topographies of dual lead-crowned gear surfaces: (a) Case A ($b = 0$) and (b) Case B ($b = -2.75 \times 10^{-3}$).

4.2. Tooth contact analysis on the dual lead-crowned tooth surface

As shown in figure 9(a), a possible application for the dual lead-crowned gears is considered in this example, where Gear 1 is the driving gear meshing with another two gears, Gear 2 and Gear 3, simultaneously. The related parameters for these three gears can be seen in table 2.

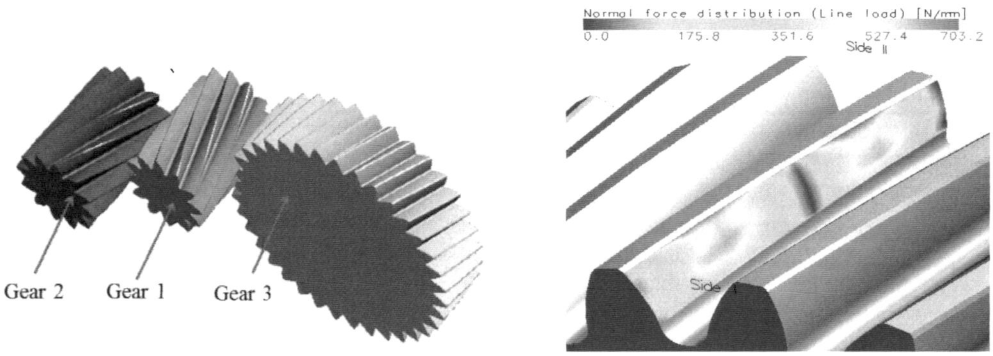

(a) (b)

Figure 9. Three-gear train and its TCA result: (a) 3-D model, (b) Normal force distribution.

The tooth contact analysis (TCA) of this gear train is conducted with the input power of 60 kW and the input rotation speed of 1000 rpm to display the effect of applying dual lead-crowning on the Gear 1 by giving $k = -3.5 \times 10^{-3}$ and $b = -2.75 \times 10^{-3}$. As the TCA result show in figure 9(b), the tooth contact pattern is obviously divided into two elliptical zones at the centers of two half tooth flank sections and no edge contact occurs on both end sides of tooth flank. Although the higher contact forces appear at the tooth tips of two half sections, but these situations occur generally on the tooth surface with or without lead crowning while the profile modification of tooth tip or root is not applied. by this example, it displays the merit of the proposed dual lead-crowing method because of no normal force concentrations on both end edges of Gear 1 tooth flank under loads, as well as on Gear 2 and Gear 3.

Table 2. Parameters for three gears.

Parameters	Gear 1	Gear 2	Gear 3
Number of teeth	12	12	36
Tooth Face-width	40.0 [mm]	40.0 [mm]	20.0 [mm]
Normal module		2.0 [mm]	
Normal pressure angle		30.0 [degree]	
Helical angle		20.0 [degree]	

5. Conclusion

A dual lead-crowning method for wide face-width gears using the internal honing machine is proposed in this paper. The mathematical models of honing wheel generation and gear honing process are also established. By controlling swivel angle of the honing stone and giving an additional rotation angle function for work gear, the dual lead-crowned and anti-twist gear tooth flanks can be achieved. Additionally, the numerical examples display the merits of the proposed method by means of topographies of generated gear tooth surfaces and TCA in a possible application for this kind of gear with dual lead-crowned tooth surfaces.

Acknowledgments

The authors are grateful to the Ministry of Science and Technology of Taiwan (R.O.C.) for the financial support under the projects No: MOST 105-2221-E-008-047.

References

[1] J P Dugas 1988 *Rotary Gear Honing, The 16th Annual AGMA Gear Symposium* pp 36–38

[2] J P Dugas 1992 *Gear Finishing by Shaving, Rolling & Honing - Part II, Gear Technology* pp 24–30

[3] N W Wright and H Schriefer 1997 *"Basic Honing and Advanced Free-Form Honing," Gear Technology, Vol.* **204** pp 26–33

[4] F Klocke 2009 *"Manufacturing Processes 2-Grinding, Honing, Lapping," RWTHedition, RWTH Aachen* pp 334–337

[5] D T Mehta and M G Rathi 2013 *"A Review an Internal Gear Honing," International Journal of Engineering Research & Technology (IJERT)* Vol. **2** Issue 5

[6] N Amini, H Westberg, F Klocke and T K¨ollner 1999 *"An Experimental Study on the Effect of Power Honing on Gear Surface Topography," Gear Technology, Vol.* **16** pp 11–18

[7] V T Tran, R H Hsu and C B Tsay 2014 *"Study on the Anti-Twist Helical Gear Tooth Flank with Longitudinal Tooth Crowning," Journal of Mechanical Design* Vol. **136** pp 061007 (1–10)

[8] V T Tran, R H Hsu and C B Tsay 2015 *"A Crowned Helical Gear with Twist-free Tooth Flanks Generated by A CNC Honing Machine," International Journal of Mechanical and Production Engineering* Vol. **3** Issue 1

[9] Y R Wu and V T Tran 2015 *"Lead Crowning and Anti-twist for Tooth Flank of a Heat Treated Helical Gear on Internal CNC Honing Machine," Applied Mechanics and Materials* Vol. **799–800** pp 554–59

[10] Y R Wu and V T Tran 2016 *"Transmission and Load Analysis for A Crowned Helical Gear Pair with Twist-free Tooth Flanks Generated by An External Gear Honing Machine," Mechanism and Machine Theory* Vol. **98** pp 36–47

[11] F L Litvin 1994 *Gear Geometry and Applied Theory, PTR Prentice Hall, Englewood Cliffs, NJ* pp 412–68

Parameter optimization of electrochemical machining process using black hole algorithm

Dinesh Singh* and Rajkamal Shukla

Mechanical Engineering Department, Sardar Vallabhbhai National Institute of Technology, Surat - 395007, India

*E-mail: dineshsinghmed@gmail.com

Abstract. Advanced machining processes are significant as higher accuracy in machined component is required in the manufacturing industries. Parameter optimization of machining processes gives optimum control to achieve the desired goals. In this paper, electrochemical machining (ECM) process is considered to evaluate the performance of the considered process using black hole algorithm (BHA). BHA considers the fundamental idea of a black hole theory and it has less operating parameters to tune. The two performance parameters, material removal rate (MRR) and overcut (OC) are considered separately to get optimum machining parameter settings using BHA. The variations of process parameters with respect to the performance parameters are reported for better and effective understanding of the considered process using single objective at a time. The results obtained using BHA are found better while compared with results of other metaheuristic algorithms, such as, genetic algorithm (GA), artificial bee colony (ABC) and bio-geography based optimization (BBO) attempted by previous researchers.

1. Introduction

Advanced machining processes (AMP) have begun a new era in the current metal age. As the traditional machining processes are not offering sustainable machining quality for the harder material, the AMPs are utilized in the modern industries as a suitable choice. The demand of advance machining process (AMP) is increased due to its capability of generating complex profiles in the machined component. The significance of AMP is that they utilize altered forms energy instead of traditional tools for machining operations. In this paper, an attempt is made to obtain optimum parameters to improve the performance of the electrochemical machining (ECM) process using Black hole algorithm (BHA). The algorithm provides optimum solution with less number of computational trials. The successful application of BHA is found in power distribution problems [1, 2].

The present paper is organized as follows. Section 2 briefly reviews the previous literature works on the considered machining processes. Section 3 presents the proposed method, BHA. Section 4 shows the application of BHA on the considered processes for obtaining the optimum solutions with results and discussion. Finally, conclusions are presented in Section 5.

2. Literature Survey

Parameter optimization of ECM process parameters, i.e., tool feed rate, applied voltage and electrolyte flow velocity have been attempted to enhance geometrical inaccuracy using real-coded genetic algorithm [3]. Similarly, parametric optimization of ECM, electrochemical discharge machining and electrochemical micromachining using artificial bee colony (ABC) were attempted and the results were compared [4]. The effect of the ECM process parameters during machining of EN-31 steel on material removal rate (*MRR)*, average roughness (*Ra*), overcut (*OC*) and cylindricity error were studied using gray relation analysis [5]. The effect of applied voltage, electrolyte concentration, electrolyte flow rate on *MRR* and *Ra* during machining of LM25 Al/10%SiC composites were reported using response surface method (RSM) [6]. The influences of contaminations on electro-erosion based machining process have been studied to increse the process accuracy in material removal [7]. Parametric optimization of ECM and wire electro chemical turning were attempted using genetic algorithm (GA), ABC and bio-geography based optimization (BBO) algorithms and found the superiority of the BBO algorithm in terms of the optimum solution and computation time [8]. A cuckoo search algorithm (COA) was attempted for parameter optimization of ECM process and the effects on *MRR* and *Ra* were reported [9]. The effect of the parameters, current, voltage, flow rate of electrolyte and inter-electrode gap on *MRR* and *Ra* were reported using RSM for hardened steel and non-dominated sorting genetic algorithm (NSGA-II) was applied to get alternative solutions [10]. An experimental investigation was reported to determine the influence of ECM parameters, like, electrolyte concentration, voltage, feed rate and inter-electrode gap on *MRR* and *Ra* for EN31 tool steel [11]. An experimentation investigation was conducted on ECM process to develop the relation between the input and output characteristics [12]. A microstructure study of surfaces have been attempted for the material steel 20MnCr5 obtained during machining using ECM to control the influence of electrolyte on the machining surfaces [13]. An analytical work on the mechanism of *MRR* in the inter electrode gap like pressure and velocity variation has been reported considering nickel based alloy as workpiece material with 'I' shaped tool [14].

One of the significant variant of ECM is electrochemical micro machining (EMM). When the ECM is used in the micrometer range to obtain complex profile, it is termed as EMM. A grey relational analysis (GRA) have been applied to optimize the process parameters like, temperature, current density and electrolyte composition in electro polishing of 316L stainless steel for enhancing the multiple performance characteristics *Ra* and passivation strength [15]. An experimental investigation have been conducted using RSM to obtain the influence of EMM parameters pulse on/off ratio, machining voltage, electrolyte concentration, tool vibration frequency and voltage frequency on *MRR* and *ROC* [16]. The survey in micro-cutting operation has been reported and suggested the key research areas that can be improved machining characteristics [17]. An advancement and research area have been proposed in mechanical micromachining include process physics, materials and microstructural effects, machine tools, design issues, software and simulation tools for future developments [18].

Past researchers have applied several optimization techniques, like GA, ABC, BBO [8], Gauss-Jordan [19]. In most of the cases, the sub-optimal solutions have been obtained. In this paper, a recently developed BHA algorithm [20] is applied for the machining parameter optimization problem. To see the considered algorithm effectiveness, the results of BHA are compared with the previous results.

3. Black hole algorithm (BHA)

The use of optimization techniques is greatly adopted by the researchers. As these techniques have a wide variety of applications and can solve complex problems effectively. Several researchers have attempted the parameter optimization of both the traditional and NTM processes using non-traditional optimization techniques. In recent years, new optimization techniques are recognized. These techniques are being used to solve the high-dimensional problems. The methodology of the considered optimization technique is presented here-in-under.

BHA [20] considers the fundamental concept of the physics "black hole". It is a population based metaheuristic algorithm. The algorithm is lean-to the particle swarm optimization (PSO) algorithm and unlike PSO; BHA has less operating parameter to tune. Like other metaheuristic algorithms, star positions are generated randomly to obtain the objective function values, i.e., candidate solution. The best solution generated in the trial is labeled as a "black hole" and rest are considered as "normal stars". This obtained "black hole" is used to update the normal star position. All the stars will attract towards the star considered as a black hole and these stars will be swallowed by it. If during the motion of normal stars towards the "black hole" founds a position having optimal value then the present black hole will be replaced with this current position of optimum value as a new black hole. Now, the new stars will be updated along with this new black hole. The swallowing of the stars by black hole is computed using Eq. (1) as given in [20]. The star position with worst fitness function value is swallowed by the black hole and new star's position are generated and updated. This process is repeated till the termination criterion satisfied.

$$x_i(t+1) = x_i(t) + rand(x_{BH} - x_i(t)) \quad i = 1,2,....,N \tag{1}$$

where $x_i(t)$ and $x_i(t+1)$ are the star position. x_{BH} is the black hole location. *rand* is a random number between 0 and 1. *N* is total population of stars.

4. Parameter optimization of ECM Process

An ECM process works on the principle of electrolysis. An electrochemical reaction occurs in the process which dissolves the reaction products produced on the workpiece material. The process resembles with reverse electroplating process. A current of electrolyte fluid carry's away the removed material before it settles down at the ground level. The hardness of the material does not affect the ECM performance which makes it suitable for machining hard materials. The primary application of ECM is to produce complicated patterns in the products made of electrically conducting but difficult-to-hard materials. Each machining process has number of process parameters and these parameters considerably affect the operation of the process. The choice of the optimum process parameters is thus become significant to enhance the performance of the process. So, an attempt is made to optimize the selected parameters, i.e., electrolyte concentration, electrolyte flow rate, applied voltage and inter-electrode gap of the considered case study.

The considered problem is taken from Bhattacharya and Sorkhel [19]. They have used an automatic microprocessor-based tool feed system for machining cylindrical steel EN-8 having dimension 19 mm in diameter [20]. They have used Sodium chloride (NaCl) as an electrolyte during machining. They considered four process parameters at level five with coded range between the values -2 to 2 to control the considered machining setup. The coded value of each process parameter can be computed using the Eq. (2). The range of the process parameters (actual values) is given as follow: Electrolyte concentration (x_1): (15 g/l, 90g/l), Electrolyte flow rate (x_2): (10 l/min, 14 l/min), Applied voltage (x_3): (10 V, 30 V) and Inter-electrode gap (x_4): (0.4 mm, 1.2 mm).

$$Coded\ Value = \frac{2x - (x_{\max} + x_{\min})}{(x_{\max} - x_{\min})/2} \tag{2}$$

where, x_{max} and x_{min} are the maximum and minimum actual values of the parameters.

Bhattacharya and Sorkhel [19] have used a central composite based second-order experimentation plan using RSM to develop the regression model for *MRR* and *OC* as given in Eqs. (3) and (4). The same regression models are considered in the present work to optimize the process parameters.

$$MRR = 0.6244 + 0.1523x_1 + 0.0404x_2 + 0.1519x_3 - 0.1169x_4 + 0.0016x_1^2 + 0.0227x_2^2 + 0.0176x_3^2 - 0.0041x_4^2 + 0.0077x_1x_2 + 0.0119x_1x_3 - 0.0203x_1x_4 + 0.0103x_2x_3 - 0.0095x_2x_4 + 0.0300x_3x_4 \tag{3}$$

$$OC = 0.3228 + 0.0214x_1 - 0.0052x_2 + 0.0164x_3 + 0.0118x_4 - 0.0041x_1^2 - 0.0122x_2^2 + 0.0027x_3^2 + 0.0034x_4^2 - 0.0059x_1x_2 - 0.0046x_1x_3 - 0.0059x_1x_4 + 0.0021x_2x_3 - 0.0053x_2x_4 - 0.0078\,x_3x_4 \tag{4}$$

4.1. Single objective optimization results using BHA algorithm

The BHA is applied to the considered ECM problem for parameter optimization. The effectiveness of the considered algorithm is measured using Eqs. (3) and (4). Here, the performance parameter *MRR* and *OC* are to maximized and minimized separately. Bhattacharya and Sorkhel [19] have attempted the considered problem using Gauss-Jordon algorithm. Similarly, these same models are attempted by Mukherjee and Chakroborty [8] using GA, ABC and BBO. The obtained results of BHA when compared with the results of Bhattacharya and Sorkhel [19] for the Gauss Jordan method, it is found that the results obtained are much better for *MRR* and *OC*. The results obtained by Mukherjee and Chakroborty [8] are sub-optimal solutions as they converge faster compared to the considered algorithm. The result obtained using BHA is far better compared to the results of Mukherjee and Chakroborty [8] for GA, ABC and BBO algorithms. The obtained result for *MRR* and *OC* using BHA algorithm are compared with the results of previous researcher and the comparison is given in table 1. The obtained optimum process parameters, i.e., electrolyte concentration, electrolyte flow rate, applied voltage, inter-electrode gap in coded form using BHA as {1.8932, 1.6466, 1.9407, -1.7416} and {-1.8548, -1.8650, -1.9132, -1.9316} for *MRR* and *OC* respectively. The values obtained in coded form are decoded (i.e. the actual values of process parameters) using Eq. (2) as depicted in table 2 (a). The actual values are rounded off so that they can be easily tuned in the considered process to achive the desired perfromance characteristics. The feasible values of the process parameter along with the coded values are depicted in table 2 (b). The consistency of the BHA result are verified with 50 trials. The mean, standard deviation and computational time (sec) are recorded as {1.6892, 0.0482, 0.5674} and {0.1305, 0.0100, 0.5974} for *MRR* and *OC* respectively. The low value of standard deviation shows the consistency and effectiveness of BHA to obtained the desired accuracy for the considered problem. The convergence results for *MRR* and *OC* obtained using BHA are depicted in figure 1.

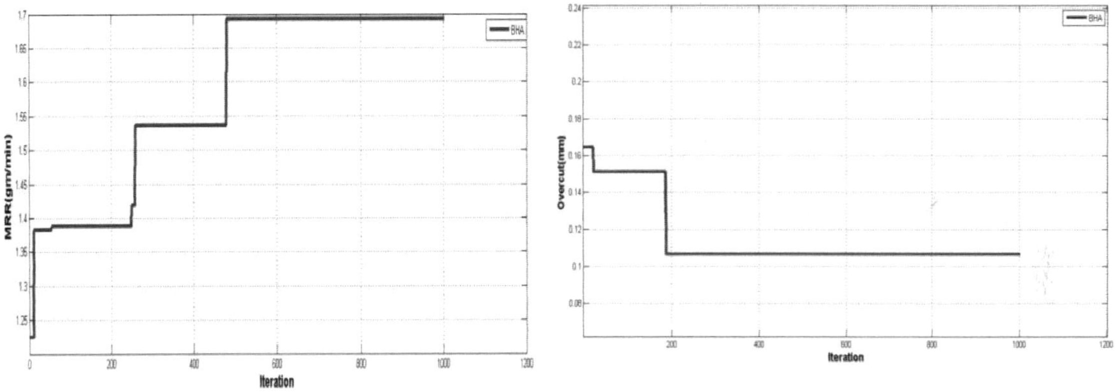

Figure 1. Performance parameter convergence using BH algorithm.

Table 1. Comparison of BHA results.

Algorithm	Gauss-Jordan [19]	GA [8]	ABC [8]	BBO [8]	BHA
***MRR* (g/min)**	0.8230	1.1603	1.3077	1.5069	1.6922
***OC*(mm)**	0.2706	0.2369	0.2067	0.1320	0.1063

Table 2 (a). Optimum values obtained using BHA.

Performance parameter	*Electrolyte concentration (g/l)*		*Electrolyte flow rate(l/min)*		*Applied voltage (V)*		*Inter-electrode gap (mm)*	
	Coded	*Actual*	*Coded*	*Actual*	*Coded*	*Actual*	*Coded*	*Actual*
***MRR* (g/min)**	1.8932	87.9975	1.6466	13.6466	1.9407	29.7035	-1.7416	0.4517
***OC* (mm)**	-1.8548	17.7225	-1.8650	10.1350	-1.9132	10.4340	-1.9316	0.4137

Table 2 (b). Round-off and feasible values of process parameters obtained using BHA.

Performance parameter	*Electrolyte concentration (g/l)*		*Electrolyte flow rate (l/min)*		*Applied voltage (V)*		*Inter-electrode gap (mm)*	
	Coded	*Actual*	*Coded*	*Actual*	*Coded*	*Actual*	*Coded*	*Actual*
MRR **(g/min)**	1.8933	88	2	14	2	30	-2	0.4
OC **(mm)**	-1.8400	18	-2	10	-2	10	-2	0.4

4.2. Effects of process parameters

The optimality of the obtained solution can be verified with the graphs trends for the process parameters with respect to *MRR* and *OC* as depicted in figures 2 (a) - (d). The optimal solution is considered as a constant and one process parameter is varied simultaneously to plot the trends. The influence of process parameters on *MRR* and *OC* are studied using graph trends. These trends are used to understand the physics of considered process in terms of effect produced by the variations of process parameters on the performance. As shown in the figures 2 (a) - (d), *MRR* increases with an increase in electrolyte concentration (figure 2 (a)), electrolyte flow rate (figure 2 (b)), applied voltage (figure. 2 (c)), but its value decreases with the increase of inter-electrode gap (figure 2 (d)). These trends of MRR with respect to process parameters are justified with an argument that increament of MRR with electrolyte concentration because increase of electrolyte flow rate and concentration cause change in increase of quantity of negative electrolytic ions to produce electrochemical reactions with the workpiece metallic ions. The increase in applied voltage has greater influence of machining current which is utilized for machining thereby causing improvement of the MRR. Therefore, the maximum possible value of the considered process parameters will be the optimum solution for MRR.

While the parameter *OC* increases with the increase of electrolyte concentration (figure 2 (a)), the value of *OC* increases with the increase of the electrolyte flow rate up to a certain limit and then gradually decreases (figure 2 (b)). *OC* value also increases with the increase of applied voltage (figure 2 (c)) and inter-electrode gap (figure 2 (d)). The increase in electrolyte flow rate causes an increment in *OC* because of the greater volume of electrolytic ions available in the machining area. Furthermore, increase of applied voltage causes a greater current. The effects of current flow weaken gradually, as the rapid removal of products and bubbles from machining zone. Moreover, an increased inter-electrode gap has a nonlinear effect on the *OC* as it weakens the current effect at the boundaries of the flow path. Therefore, minimum values of process parameters should be used to enhance *OC*. These trends of the performance parameters *MRR* and *OC* confirm the optimality of the solution obtained using BHA.

5. Conclusions

For the considered ECM process, the MRR obtained using BHA is increased from 0.8230 g/min to 1.6922 g/min which is very significant improvement and the performance parameter OC is decreased from 0.2706 mm to 0.1063 mm. So, the considered algorithm is giving satisfying results to get the optimum parameter setting on ECM process. Furthermore, the consider BHA is providing the optimum solution with less number of trials and less computational time. It shows the effectiveness and applicability of BHA to the parameter optimization of ECM process.

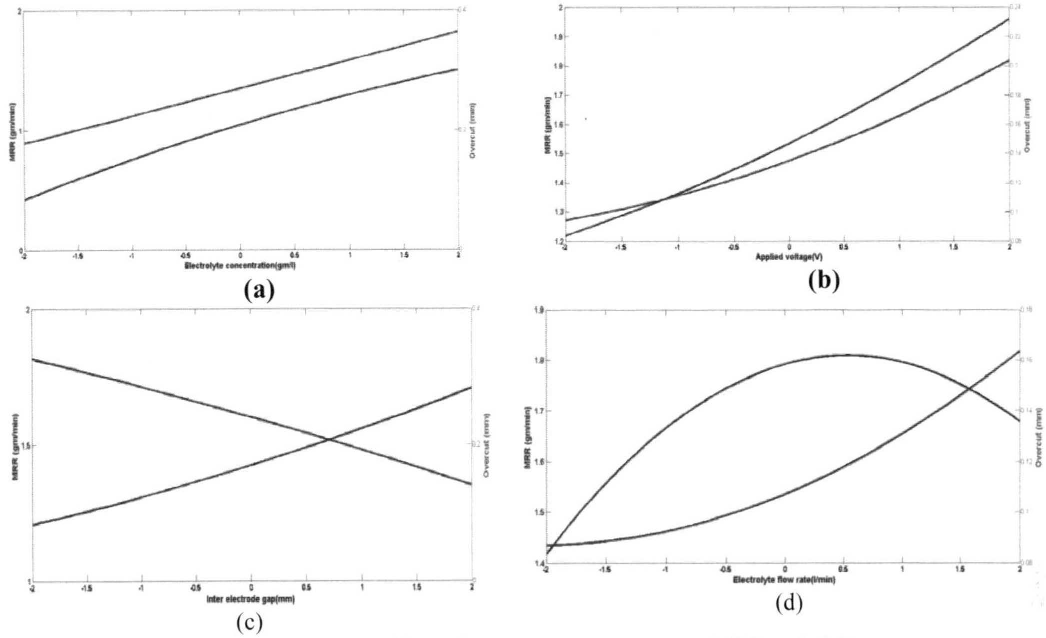

Figure 2. Effect of process parameter on *MRR* and *OC*.

References

[1] K. Lenin, B.R. Reddy, M.S. Kalavathi, Int. J. Res. Manag. Sci. Technol. **2,** 10–15, (2014)
[2] R. Azizipanah-abarghooee, T. Niknam, F. Bavafa, M. Zare, Electr. Power Syst. Res. **108**, 16–34, (2014)
[3] N.K. Jain, V.K. Jain, Mach. Sci. Technol. **11**, 235–58, (2007)
[4] S. Samanta, S. Chakraborty, Engi. Appl. Artif. Intel. **24**, 946-57 (2011)
[5] D. Chakradhar, A.V. Gopal, Int. J. Model. Opt. **1**, 113–17, (2011)
[6] C. Senthilkumar, G. Ganesan, R. Karthikeyan, Transac. Nonferrous Metals Soc. China. **21**, 2294–2300, (2011)
[7] H.P. Schulze, W. Schatzing, Procedia CIRP. **6**, 58-63, (2013)
[8] R. Mukherjee, S. Chakraborty, Int. J. Adv. Manuf. Technol. **64**, 781-91, (2013)
[9] R. Teimouri, H. Sohrabpoor, Front.Mech. Engi. **8**, 429–42, (2013)
[10] B.R. Acharya, C.P. Mohanty, S.S. Mahapatra, Procedia Engi. **51**, 554–60, (2013)
[11] M.K. Das, K. Kumar, T.K. Barman, P. Sahoo, Procedia Mater. Sci. **6**, 729–40, (2014)
[12] H. Sohrabpoor, S.P. Khanghah, S. Shahraki, R. Teimouri, Int. J. Adv. Manuf. Technol. **82**, 1683-92, (2015).
[13] S. Ayyappan, K. Sivakumar, Int. J. Adv. Manuf. Technol. **82**, 2053-64, (2015).
[14] A.A. Nayak, S. Gangopadhyay, D.K. Sahoo, J. Manuf. Process. **23**, 269–77, (2016)
[15] P. S. Kao, H. Hocheng, J. Mater. Process. Technol, **140**, 255–59, (2003)
[16] J. Munda, B. Bhattacharyya, Int. J. Adv. Manuf. Technol. **35**, 821-32, (2008)
[17] J. Chae, S.S. Park, T. Freiheit, Int. J. Mach. Tools and Manuf. **46**, 313-32, (2006)
[18] D. Dornfeld, S. Min, Y. Takeuchi, CIRP Annals – Manuf.Technol. **55**, 745-68, (2006)
[19] B. Bhattacharyya, S.K. Sorkhel, J. Mater. Process. Technol. **86**, 200-07, (1999)
[20] Hatamlou, Infor. Sci. **222**, 175–84, (2013)

IOP Conf. Series: Materials Science and Engineering **282** (2017) 012007 doi:10.1088/1757-899X/282/1/012007

Failure investigations of failed valve plug SS410 steel due to cracking

V. D. Kalyankar* and D. D. Deshmukh

Mechanical Engineering Department, S. V. National Institute of Technology, Surat, Gujarat, India

*E-mail: vivekkalyankar@yahoo.co.in

Abstract. Premature and sudden in service failure of a valve plug due to crack formation, applied in power plant has been investigated. The plug was tempered and heat treated, the crack was originated at centre, developed along the axis and propagates radially towards outer surface of plug. The expected life of the component is 10-15 years while, the component had failed just after the installation that is, within 3 months of its service. No corrosion products were observed on the crack interface and on the failed surface; hence, causes of corrosion failure are neglected. This plug of level separator control valve, is welded to the stem by means of plasma-transferred arc welding and as there is no crack observed at the welding zone, the failure due to welding residual stresses are also neglected. The failed component discloses exposed surface of a crack interface that originated from centre and propagates radially. The micro-structural observation, hardness testing, and visual observation are carried out of the specimen prepared from the failed section and base portion. The microstructure from the cracked interface showed severe carbide formation along the grain boundaries. From the microstructural analysis of the failed sample, it is observed that there is a formation of acicular carbides along the grain boundaries due to improper tempering heat treatment.

1. Introduction

Failure of mechanical components is an unpreventable issue in power, chemical, refinery and nuclear industries due to severe working conditions such as high-pressure steam, corrosion, fatigue, wear and abrasion etc. The components of control valve, turbine, boiler and pump such as, stem, pipes, shafts, gears, impellers, tubes, flanges, couplings etc. fails regularly during the operation. During ongoing operation, a shutdown for fixing these damages costs a lot; it is venerable to carry out prior inspections on the handy units to avoid this kind of failure. In general, power industries involve very large industrial complexes containing various units. The common causes of the failure of mechanical components used in power, chemical, refinery and nuclear industries are investigated earlier by researchers that are, stress corrosion cracking, welding failure due to fatigue, pitting corrosion, residual stresses, improper heat treatment of components etc. Delavar et al. [1] investigated the causes of cracking of stainless steel flange used in petroleum refining and the petrochemical operations. It was investigated that the failures occurred due to stress corrosion cracking caused by the presence of chloride in the anti-seize grease due to high temperatures and high-pressure condition. Roy et al. [2] investigated the in-service failure of an economizer tube and it was observed that, a localised corrosion directed towards the failure of economiser tubes forming corrosive products and cracks. Suess et al. [3] carried out the investigation of in-service cracking of turbine inlet block made up of 410 stainless steel.

the investigations was carried out with scanning electron microscope (SEM) and it was revealed that, the crack was developed due to formation of grain boundary carbides and these carbides likely formed during tempering of the part. Schroeder [4] investigated the metallurgical failure analysis of a fractured steam control valve stem made up of Incoloy 901 alloy. SEM with energy dispersive x-ray spectroscopy, tensile testing, hardness testing, chemical analysis and metallographic analysis was carried out to trace the exact reason of failure. Jung et al. [5] carried out the metallographic analysis of the specimen prepared from failed stem of isolation valve used in nuclear power plant. The identified causes of failure was precipitation of non-metallic inclusions with higher hardness at the failed region, this was due to improper tempering process applied to stem. Parnian [6] carried out the failure investigations of austenitic stainless steel tubes used in a gas fired steam heater. The cracks developed on the tube were examined and hardness was measured at the seam weld, heat affected zone and U-bend areas. Microstructure investigations were carried out with optical microscopy and SEM and it was revealed that the cause of cracking was stress corrosion. Das et al. [7] investigated the premature failure of the impeller blade of coke plant, which reveals exposed surface of a crack. The microstructural observation at weld zone was carried out and it showed severe intergranular corrosion, which causes the failure. Shirinzadeh-Dastgiri [8] had carried out the failure investigations of failed weld joint in AISI 1518 low carbon steel pipeline with the help of non-destructive testing, radiographic and metallographic analysis. The identified causes of failure was improper welding since, micro-cracks were noticed on the weld interface and possible corrosion attack due to cracking, this corrosion mechanism was studied by electrochemical techniques. Sharma and Roy [9] investigated the failure of AISI stainless steel pointer rod, visual inspections, microstructural investigations with optical and SEM were carried out. It was investigated that, the presence of MnS inclusions and deformation-induced martensite affected the pitting corrosion resistance causes the failure of pointer rod. Qayyum et al. [10] carried out numerical simulation of cracked disc of AISI H-11 tool steel, results showed significant drop in hoop stress with increase in number of cracks, thus limiting the number of cracks possible in a thermal fatigue crack network. Tsai et al. [11] studied the phase transformation phenomenon in AISI 410 steels that will hepled in detrmining the causes of failure. Raghu Shant et al. [12] carried out the failure analysis of cracks formed at extrados of bend pipe. Recently Ma et al. [13] had carried out the failure investigations of circulating water pump of duplex stainless steel of 1000 MW ultra-supercritical thermal power unit.

As, remaining life assessment (RLA) of aged components in petrochemical in power industry becoming very important for economy and safety reasons [2]. Root cause analysis need to undertake to investigate the failure of a mechanical components to avoid such future failures. This paper describes a case history of the failure investigation of damage valve plug used in separator valve of steam power plant.

2. Case history

Control valves are the main mechanical components used to control the flow of fluid, specified volume of the flow to a specified pressure level. The basic components of the valve are stem, plug, the bearing housing and the casing. For the proper controlling of the valve, all parts must remain in the proper functioning and working condition in order to insure safe working. A premature failure of valve plug used in separator level control valve connected to steam blowing pipes due to cracking was reported on 22nd Sep 2015. Before in service, the plug was tempered and heat-treated as per requirement of the user and application. The component operated for the opening of steam flow at a pressure of 16.58 Mpa and the temperature of steam about 450-550^{0}C. The expected life of the component is 10-15 years while, the component had failed just after the installation that is, within the 3 months of its service. The details site images of crack generation and failure is as shown in Figure 1. After reviewing the damaged trim pictures of failed valve from site location, it was anticipated that this could be due to delayed thermal cracking which may be due to heat treatment process tempering temperature cycles and its associated soaking time. Material for plug (SS410) is a heat-treated and is a subcontracted activity, where as per the approved procedure these activities are being carried out.

From the initial investigation, it was assumed that, the tempering might had been created problem in the HT process, where the quick cooling might have left some thermal stresses within the core of the material (Not on the surface) and after a period of time lying idle this plug might have released those stresses causing cracks axially. It was required to study the nature and likely cause of cracking. To investigate the exact issue of the crack, the methodology explained in next section is considered for the further investigations.

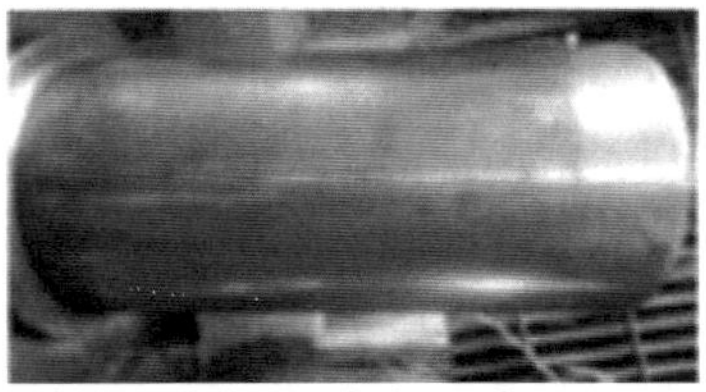

Figure 1. Site failure of plug of separator valve

3. Methodology

From the previous case studies of failure investigations in the past, it is found that the common methodology adopted by the investigator consist of, visual examination, radiographic examination, metallographic examination, energy dispersive spectroscopy and SEM etc. [1-8]. These approaches are useful to trace the exact causes of the failure. To identify the exact causes of failure, in the present study following approaches are used as a methodology, in which the evaluation of the main causes of failure can be identified, that consist of:

- Determination of material's properties, such as yield, tensile strength, hardness and chemical composition.
- Visual analysis of crack surfaces.
- Metallurgical analysis of base material and cracked (failed) zone.

After receiving the cracked plug from the site, from visual observation it is perceived that the plug cracked along the axial direction, which was propagated form the centre and towards the radial direction as shown in Figure 2 (a) and 2 (b). For the detailed analysis, a plug is sliced at the bottom section and disc shape portion is removed which is followed by visual inspection, hardness and chemical testing on the disc portion, etc. Two square blocks from the disc were removed, one from the crack interface containing crack and one which is away from the crack interface representing the base material. The block that is removed from the crack section directly exposes the cracked surface after opening, which was automatically separated out and crack interface turn out to be open for observations. After the crack opening, no corrosion product was seen by visual inspection of the plug at crack interface; also at weld interface, no crack was observed. So the causes of failure due to corrosion and welding residual stresses are neglected for the investigations. Hardness of the failed component is measured and chemical composition is evaluated by the Atomic Emission Spectrometry technique. Microstructural observations of the failed region is also carried using optical microscope according to ASTM E 407.

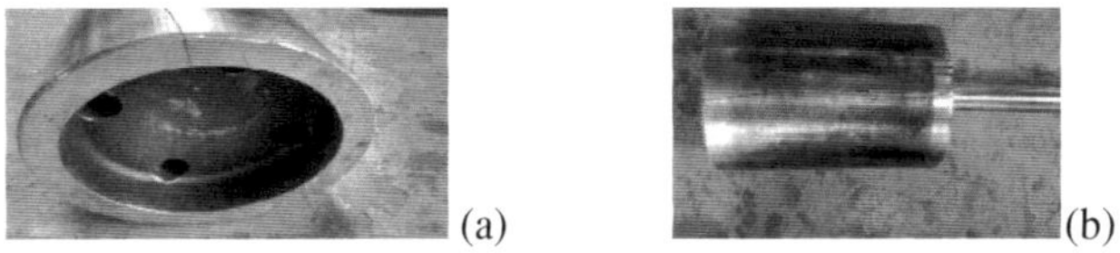

(a) Failed plug-showing crack propagating along radial direction
(b) Crack propagating axially
Figure 2. (a) and (b) Failed plug-showing crack

3.1 Chemical composition of the part

The chemical composition at the crack interface and other location was measured on the disc shaped sliced portion. Chemical composition obtained through spectrometer is shown in Table 1. The chemical composition is found consistent throughout the slice surface which meets the specified requirement as per type of 410 stainless steel. This interface composition also matches with the composition reported by manufacturer, before the failure of component.

Table 1. Chemical composition of plug (Wt %)

C	Cr	Mn	Si	S	P	Fe
0.1	11.7	0.8	0.6	0.15	0.2	87

3.2 Hardness testing

Hardness testing was carried out at various location on disc shaped sliced portion to evaluate the change in the properties of material. It is observed that, at the cracked interface, higher hardness was reached. The measured hardness near the cracked interface was of 44-46 HRC and that of base material it was 39-42 HRC. The hardness values measured at various points on the sliced sample from the failed plug is shown in Figure 3.

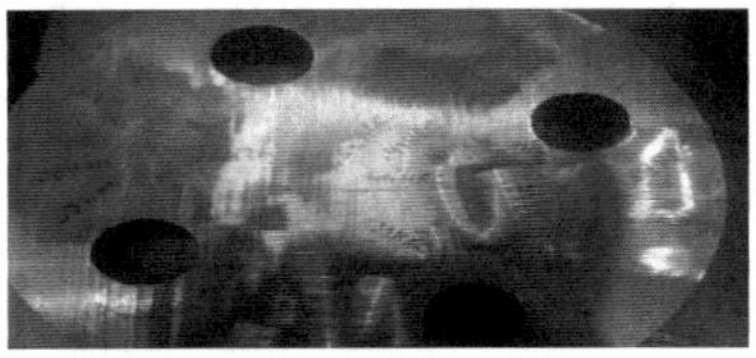

Figure 3. Hardness at various zones of sliced section

The specimen along with image view of the crack prior to and after opening is shown in Figure 4 (a) and (b). The opened crack exhibits features that indicate relatively brittle fracture. The measured hardness of the base material varies in the range of 39-42 HRC is consistent with the quenched and tempered condition of material but at cracked portion higher hardness is noticed than other locations.

(a) (b)

Figure 4. Block from cracked portion showing open crack

3.3 Metallographic analysis

For the metallographic study, from each sectioned block, one small cross section sample is removed. One from the cracked interface and other away from cracked to represent base material and to document the crack morphology using optical microscope.

The main crack surface shown in Figure 4 (b) shows brittle failure zone. The cracks started from the inner surface, i.e. from the centre of plug, which travels axially and propagated along the radius, towards to the outer surface. In order to observe the surface microstructure, the sectioned sample was first mechanically wet ground using a 1500 grit silicon carbide (SiC) paper, then polished with aluminium oxide (Al_2O_3) powder of 3 and 0.25 mm diameter, followed by electro etching according to ASTM E 407.

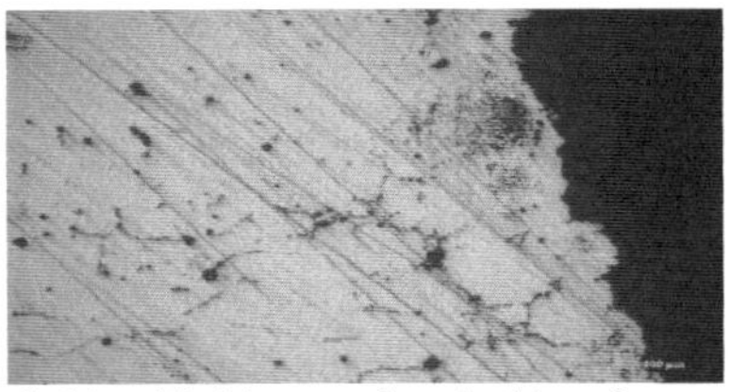

Figure 5. Magnified image of cracked interface surface

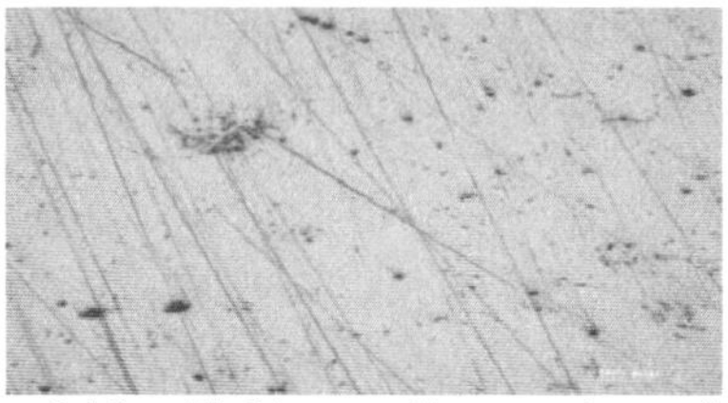

Figure 6. Magnified image of base portion surface

After etching, the specimen was cleaned with distilled water and methanol, and then dried in air. Subsequently, the specimen was examined with an inverted optical microscopy (OM). It was observed that the surface microstructure of the plug had a tempered martensitic structure, as shown in Figure 5.

The metallography of the cracked sample revealed highly distributed grain boundary carbides, which are both intergranular. Un-etched magnified image of cracked surface as shown in Figure 5, reveals various cracks originating from the cracked zone. Several abnormalities in the cracked interface region are observed in the microstructure. Micro cracks and cracks have been observed at several locations, mostly originating from the main crack. However, the microstructure from the base region that was unaffected shows a martensitic matrix without any irregularity, the magnified surface of base region without etchant is also shown in Figure 6 which does not show any such crack initiations or other effects.

Higher magnification study of the microstructure, as shown in Figure 7, revealed evidence of continuous grain boundary carbides which are mostly originated from the cracked surface. The distribution of heavy carbides along the grain boundaries can be revealed clearly in Figure 7.

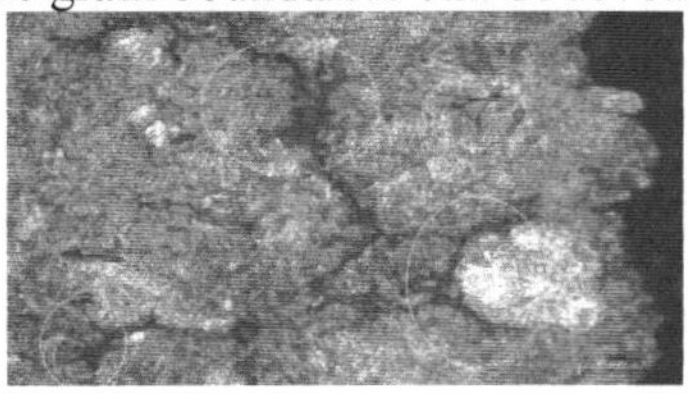

Figure 7. Microstructure of crack interface

4. Observations and Discussions

The plug used for separator level control valve was reported failed during service due to cracking. Tempering and heat treatment are the subcontracted activities were carried out on the part before installation. To carry out the failure investigations of the plug, visual inspection, chemical composition testing, hardness testing and metallographic analysis was carried out. It is observed that, the chemical composition of the part is consistent with the 410 martensitic stainless steel as per the requirement. After, the visual inspection of the cracked surface, the brittle transition nature of failure was observed. As, no corrosion products were observed on the surface of cracked interface it can be concluded that this failure did not associated with the corrosion. While, the hardness at the various location on the failed part was measured and it was observed that higher hardness was obtained at the vicinity of cracked surface than other locations. Microstructure of the cracked interface revealed heavy grain boundary carbides at the cracked interface, which are undesirable due to their brittle nature. Such carbides tend to serve as crack propagation sites, thereby significantly reducing the fracture toughness of the material. As far as the metallographic study is concerned, the carbide like structure near the grain boundaries of the cracked interface and the microhardness of the structure was higher at failed interface. That may be caused due to improper tempering of the part forming brittle failure that lacking in tensile strength resulting weaker areas along the axis and failure might occurs due to sudden impact of steam during working.

This evaluation exposed extensive indication of embrittlement at the cracked region which is due to the presence of grain boundary carbides and these carbides likely formed during tempering of the part. Based on these findings, it is recommended that the heat treating process need to be modified in order to avoid the formation of grain boundary carbides. Generally, a range of tempering temperature in between 1,050 to 1,125 °F is recommended for 410 SS steel part to avoid carbide formation [3].

5. Conclusions

The metallographic observations and hardness measurements of the specimen prepared from the failed portion lead to the conclusion that, the heat treatment applied to the valve plug might have gone wrong and the tempering caused the failure. The microstructure showing the continuous grain boundary carbides that leads to weaken the areas for tensile strength and cracks may be formed due to fatigue and excessive pressure during the installation and that might cause failure. The hardness values measured near the fracture and in the vicinity of the crack were all greater than 44 HRC. This conclusion is corroborated by comparing it to the hardness of base material which is 39-42 HRC of the plug, which is the magnitude of hardness typically found in tempered 410 martensitic stainless steels using recommended heat treatment practices. But, as at the failed interface, the higher hardness was attained with heavy carbides along the grain boundary. This was attributed due to the improper tempering process of the part. In the present investigations, the attempt is made to determine the type of failure that occurred; the problem appears to involve an improper heat treatment as per the metallographic analysis. Furthermore, Finite element analysis (FEM), SEM and 'X' ray diffraction analysis (XRD) can be performed to evaluate and explore the detail causes of failure.

6. References

[1] A N Delavar, M Shayegani and A Pasha 2013 *An investigation of cracking causes in an outlet RTJ flange in ISOMAX unit, Case Stud. Eng. Fail. Anal.* **1** pp 61–66
[2] H Roy, P Sharma, D Ghosh and A K Shukla 2014 *Corrosion failure of an in-service economizer tube J Fail. Anal. and Preven.* **14** pp 454–58
[3] S J Suess and P E 2007 *Failure analysis of a cracked turbine inlet block, Microsc. Microanal.* **13**
[4] C J Schroeder 2015 *Metallurgical failure analysis of a fractured steam control valve stem, J Fail. Anal. and Preven.* **15** pp 370–78
[5] Y C Jung, T H Lee and M T Kim 2013 *Failure analysis of isolation valve stems in steam generator blow down system, Eng. Fail. Anal.* **32** pp 98–105
[6] N Parnian 2012 Failure analysis of austenitic stainless steel tubes in a gas fired steam heater, Matr, and Des. **36** pp 788–79
[7] S Das, G Mukhopadhyay, S Bhattacharyya 2014 *Failure analysis of the impellers of coke plant, Case Stud. Eng. Fail. Anal.* **2** pp 157–61
[8] M Shirinzadeh-Dastgiri, J Mohammadi, Y Behnamian, A Eghlimi and A Mostafaei 2015 *Metallurgical investigations and corrosion behavior of failed weld joint in AISI 1518 low carbon steel pipeline, Eng. Fail. Ana.* **53** pp78–96
[9] P Sharma and H Roy 2014 *Pitting corrosion failure of an AISI stainless steel pointer rod, Eng. Fail. Anal.* **44** pp 400–07
[10] F Qayyum, M Shah, O Shakeel, F Mehdi Salem, F Rezai-Aria 2016 *Numerical simulation of thermal fatigue behavior in a cracked disc of AISI H-11 tool steel Eng. Fail. Anal.* **62** pp 242–53
[11] M C Tsai, C S Chiou, J S Du and J R Yang 2002 *Phase transformation in AISI 410 stainless steel, Mater. Sci. Eng.* A **332** pp 1–10
[12] J Raghu Shant, B K Pandey, R K Goyal and T S Kathayat 2013 *Failure analysis of cracks formed at extrados of bend pipe, of API 5L X65M Grade*, J Fail Anal and Preven. **13** pp 531–37
[13] Y Ma, S Yan, Z Yang, G Qi and X He 2015 *Failure analysis on circulating water pump of duplex stainless, steel in 1000 MW ultra-supercritical thermal power unit, Eng. Fail. Anal.* **47** pp 162–77

IOP Conf. Series: Materials Science and Engineering **282** (2017) 012008 doi:10.1088/1757-899X/282/1/012008

Thermodynamic analysis of steam-injected advanced gas turbine cycles

Devendra Pandey and Mukund H. Bade*

Mechanical Engineering Department, Sardar Vallabhbhai National Institute of Technology, Surat Gujarat 395007, INDIA

*Corresponding author: bmh@med.svnit.ac.in, mukundbade @gmail.com

Abstract. This paper deals with thermodynamic analysis of steam-injected gas turbine (STIGT) cycle. To analyse the thermodynamic performance of steam-injected gas turbine (STIGT) cycles, a methodology based on pinch analysis is proposed. This graphical methodology is a systematic approach proposed for a selection of gas turbine with steam injection. The developed graphs are useful for selection of steam-injected gas turbine (STIGT) for optimal operation of it and helps designer to take appropriate decision. The selection of steam-injected gas turbine (STIGT) cycle can be done either at minimum steam ratio (ratio of mass flow rate of steam to air) with maximum efficiency or at maximum steam ratio with maximum net work conditions based on the objective of plants designer. Operating the steam injection based advanced gas turbine plant at minimum steam ratio improves efficiency, resulting in reduction of pollution caused by the emission of flue gases. On the other hand, operating plant at maximum steam ratio can result in maximum work output and hence higher available power.

1. Introduction

The biggest challenges of today are the environmental concern due to pollution mainly flue gases, depletion of fossil fuel reserves and fluctuating prices of fuels. These challenges have motivated researchers to find new energy sources on one side and to improve existing plant efficiency on another. Gas turbine power plant mainly uses natural gas having more clean combustion but with reduced efficiency. To get the advantage of clean combustion, researchers tried to improve its efficiency by modifying basic cycle operation. Out of various advanced gas turbine cycles, in STIGT cycle, steam is formed from water by recovering heat available in exhaust of gas turbine. This heat is injected in to combustion chamber in the form of saturated or superheated steam, further improving the gas turbine cycle efficiency. This will definitely helpful for cost feasibility, if it is designed at optimal condition. Improved efficiency of STIGT will reduce pollution due to flue gas decreasing global warming effects.

A lots of work carried out on cogeneration plants [1] [2,3] [4] [5], combined cycle plants [6–10] and advanced gas turbine cycle [1] [11–13] to improve the cycle efficiency for various operating conditions. Burnham et al. [14] performed full scale test of a steam injection system and analyzed economics of plant. Foster-pegg [15] had compared turbo-charged STIGT with cogeneration plant and indicated more suitability of first one. Bolland, and Stadaas [16] evaluated water, steam and recuperative gas turbine cycle performance and compared with combined cycle. Heppenstall [2] critically reviewed advanced gas turbine cycles indicating combined cycle will be preferred for new plants. Najjar [1] had reviewed work carried by his team regarding the advanced gas turbine cycles. Horlock [17] in his book described most of the aspects of gas turbine power plant with advanced cycles. Further reduction of pollution, researchers tried to integrate gas turbine with other energy sources such as solar dish micro combined cycle [18], CO_2 capture [19,20] etc. Bade and Bandyopadhyay [21] had proposed methodology to select efficient cogeneration system based on novel graphs pressure ratio versus R curve (heat to work ratio) at maximum performance indicators. This is useful for both new as well as retrofitted plants. Fabenle et al. [22] analyzed biogas-fired integrated gasification STIGT plant based on exergy criteria. Primary objective of this paper is to propose a methodology for appropriate integration of advanced gas turbine cycle so that overall fuel consumption is reduced.

2. Steam-injected gas turbine (STIGT)

Figure 1 presents the schematic of STIGT plant with details of operation of it. Inlet air of mass flow rate m_a (at ambient temperature, T_o and pressure, P_o) is compressed until pressure P_{aco} (with a corresponding temperature of T_{aco}) is reached. Then, steam (m_s) is heated in regenerator up to temperature T_{sicc} using exhaust flue gas mixture from turbine gas. Subsequently, preheated steam (T_{sicc}) and compressed air (T_{aco}) are injected in to combustion chamber where fuel of mass flow rate m_f is burnt. The product of combustion termed as flue gas mixture (mass flow rate, m_g+m_s) at a specified maximum temperature (T_{gti}) is expanded in gas turbine to generate net work W_{net}. The flue gas at gas turbine exit temperature (T_{gto}) is supplying heat to covert water into steam. This steam is subsequently supplied to combustion chamber. Maximum ratio of mass flow rate of steam to air is accounted such that water is heated up to saturated steam condition and heat required per unit mass of air in flue gas is always greater than this required heat duty (Q_{steam}).

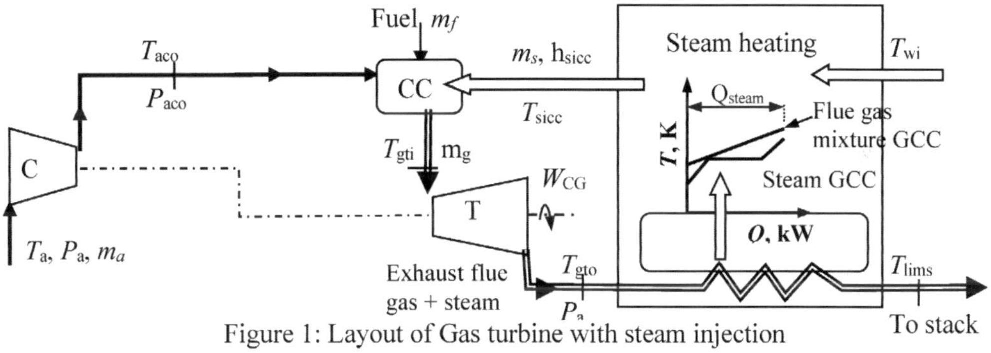

Figure 1: Layout of Gas turbine with steam injection

3. Problem definition

For a given gas turbine plant, water or steam heating requirement is known. The design data of gas turbine such as ambient air temperature (T_o) and pressure (P_o), gas turbine temperature at inlet to turbine (T_{gti}), specific heats of air and flue gas, water and steam enthalpies at required conditions, etc. are given. The minimum temperature difference between water-steam and flue gas mixture is considered 50 K. The STIGT plant supplies heat available in flue gas mixture at gas turbine exhaust to satisfy required heat duty to convert water into steam (Q_{steam}). It is assumed that the heat available in the exhaust flue gas is always greater than or equal to the required heat duty to convert water into saturated steam (Q_{steam}). This is accomplished by maintaining appropriate steam ratio S_r (ratio of mass flow rate of steam to air). The objective is to develop a methodology for optimal performance of the advanced gas turbine cycle for maximum efficiency or power output.

4. Thermodynamics analysis of STIGT

An analytical methodology for energy integration of gas turbine and steam heating, together carried out in steam injected gas turbine (STIGT) plant is developed in this section. Methodology proposed by Bade and Bandyopadhyay [21] for energy integration of gas turbine with overall process is appropriately modified to apply for STIGT for various criteria.

The heat available in flue gas mixture (gas and steam) at the outlet of gas turbine utilized to convert water into steam. The analysis is carried out per unit mass of air and important parameter considered are ratio of mass flow rate of steam to air ($S_r = m_s/m_a$) and pressure ratio of turbine cycle. Consider the gas turbine cycle plant without injection of steam (S_r=0). By using simple analysis, efficiency can be determined. If small amount of steam per kg of air is added by heating it using flue gas mixture, total mass of flue gas mixture will increase by additional steam amount. In effect, partially, extra heat energy needs to be supplied with surplus work output from gas turbine. Energy required for heating water can be presented in the form of grand composite curve (GCC) called as steam GCC [21]. A steam ratio, S_r can be targeted graphically by matching flue gas mixture line drawn from temperature of flue gas mixture at gas turbine exit with steam GCC [21], as shown in Figure 2. The ratio of mass flow rate of flue gas to steam is given by following equation:

$$\left(\frac{m_g}{m_s}\right) = \frac{(q_{steam}-q_k)-(h_{sgto}-h_{sk})}{C_{pg}\left(T_{gto}-T_k-\frac{\Delta T}{2}\right)} \tag{1}$$

where, q_k and T_k are coordinates of one of the vertexes of steam GCC with additional stack temperature and C_{pg} is specific heat capacity at constant pressure of flue gas, h represent enthalpy of steam at respective temperatures. The maximum among the flue gas flow rate determined at each vertex is the minimal flow rate of flue gas for which complete process heat demand is satisfied. At this condition, steam is at saturated condition. The point at which flue gas line just touches process GCC or stack temperature is known as utility pinch (point 5 in Figure 2) and temperature at that point termed as utility pinch temperature (T_k). Overall mass balance is written as:

$$m_a + m_f + m_s = m_g + m_s \tag{2}$$

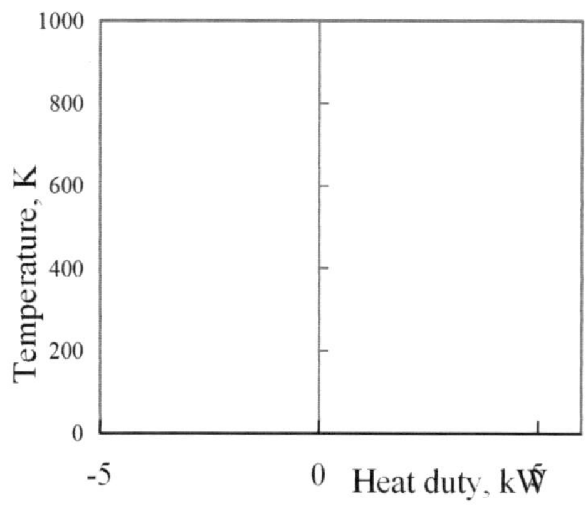

Figure 2. Targeting of S_r by matching steam heating

The compressed air (at temperature, T_{aco}, m_a) and steam are used for burning fuel in the combustion chamber. The flue gas temperature at exit of combustion chamber is T_{gti}. The energy balance for combustion chamber is given as:

$$m_a C_{pa}(T_{aco}-T_a) + m_s(h_{sicc}) + CV^*\eta_{cc} m_f = m_g C_{pg}(T_{gti}-T_a) + m_s(h_{sgti}) \tag{3}$$

where, CV is net calorific value of fuel, η_{cc} is combustion efficiency and C_{pa} is specific heat capacity at constant pressure of air.

Pinch analysis may be applied to determine steam ratio, S_r for various conditions of integration. Figure 3 illustrates the heat duty available (temperature of flue gas from gas turbine exit, T_{gto} to stack, T_{ST}) and heating required (temperature of water-steam from ambient temperature to maximum combustion chamber inlet temperature) by composite curves. The maximum superheat steam temperature possible is flue gas exhaust temperature with minimum approach temperature i.e. (T_{gto} - ΔT). Similar to GCC in pinch analysis, the GCC for flue gas mixture is a plot of enthalpy difference between utility composites as shown in Figure 3. It should be noted that the GCC for flue gas mixture is shown as reflected (see Figure 2 and 3). This helps in matching it with GCC of steam to balance utility requirement as shown in Figure 3. The matched GCC of flue gas mixture and GCC of steam, pinch point, stack losses, etc. are shown in Figure 3. All GCCs are plotted on shifted temperature scale with minimum approach temperature, ΔT. Heat lost by flue gas mixture is equal to heat received by water required to convert it into steam for minimum approach temperature. This energy balance equation can be written to get minimum S_r as:

$$\left(\frac{m_g}{m_s}\right)_{min} = \frac{\left[q_{s(Tgto-\Delta T)} - q_k\right] - \left(h_{sgto} - h_{sk}\right)}{C_{pg}\left(T_{gto} - T_k - \frac{\Delta T}{2}\right)} \tag{4}$$

where, $q_{s(Tgto-\Delta T)}$ is enthalpy of steam inlet to combustion chamber at (T_{gto} - ΔT). The minimum ratio of mass flow rate of steam to air is determined at optimal integration using Equations 2, 3, and 4 appropriately as follows:

$$S_r = \left(\frac{m_s}{m_a}\right)_{min} = \frac{\left[C_{pa}(T_{aco} - T_a) - CV\eta_{cc}\right]}{\left[\frac{(h_{sicc} - h_{wi}) - (q_s - q_k)}{C_{pg}\left(T_{gto} - T_{ig} - \frac{\Delta T}{2}\right)}\right]\left[C_{pg}(T_{gti} - T_a) - CV\eta_{cc}\right] + \left(h_{sgti} - h_{sicc}\right)} \tag{5}$$

where, q_k and T_{ig} are the coordinates of one of the vertexes of steam GCC with additional of stack temperature. q_s is total heat required for steam heating per unit mass of air and equals to the enthalpy of the superheat steam at gas turbine outlet temperature with minimum approach temperature. However, the pinch points are not known a *priori*. Since pinch is going to form at one of the vertexes of GCC of steam including stack temperature, steam rate (S_r) can be calculated following Equation 5, for each vertex (i = 1,...,k). The maximum of steam ratio (S_r) calculated for each vertex including stack temperature represents the minimum steam ratio, S_r for optimal integration of STIGT cycle. The GCC of flue gas mixture touched at any of the vertex without intersection with GCC of steam forms optimal integration of STIGT cycle, which is represented by pinch point. The other parameters w_{net}, heat supplied, and cycle efficiency are determined as appropriately per unit mass of air.

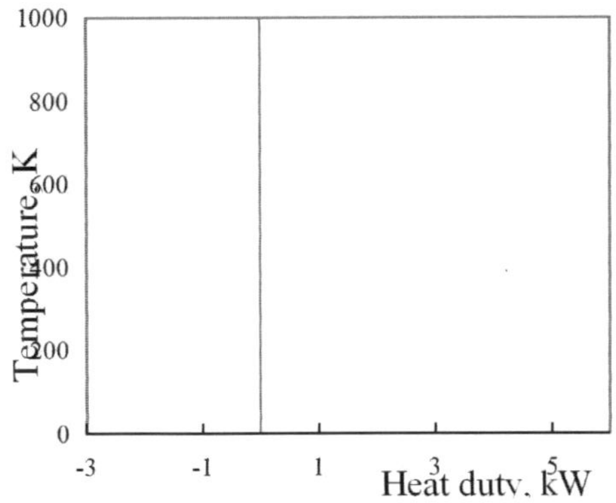

Figure 3. Targeting S_r at maximum superheat temperature

Similar to this, if steam is heated up to saturated temperature, cycle integration condition is at minimum superheat temperature, however, steam ratio, S_r is having maximum value. This integration is presented for maximum steam rate condition and determined using Equation 1 to 4 but for saturated steam condition. The proposed methodology is explained with the help of example in next section.

5. Illustrative example:

Consider advanced gas turbine cycle STIGT with design data given in Table 1. Note that all calculations are carried out per unit mass of air with steam rate varied to illustrate the proposed methodology. The minimum approach temperature for GCC of steam and flue gas mixture is 50 K.

Table 1. Gas turbine design data

Gas turbine inlet temperature (T_{gti}), K	1273	Minimum approach temperature (ΔT) K	50
Ambient temperature (T_a), K	303	Combustion efficiency (η_{cc})	0.98

Gas turbine efficiency (η_{gt})	0.9	Compressor efficiency (η_c)	0.86
Specific heat capacity of gas at const pressure (C_{pg}), kW/kgK	1.148	Sp. heat capacity of air at const pressure (C_{pa}), kW/kgK	1.005
Ratio of specific heat capacities gas (γ_g)	1.315	Ratio of specific heat capacities air (γ_a)	1.4
Net calorific value of fuel (CV), MJ/kg	42.3	Enthalpy of steam at T_{ST} (h_{sST}), kJ/kg	2716
Water inlet temperature (T_{wi}),K	298	Stack temperature (T_{ST}),K	393

Using the Equation 1 to 4 for saturated steam condition, maximum steam ratio S_{rmax} is determined for various pressure ratios. Similarly, using Equations 2 to 5, minimum steam ratio S_{rmin} is determined for the same pressure ratios. The graph, pressure ratio (PR) Vs. steam ratio is plotted for minimum and maximum steam ratios as shown in Figure 4. Maximum superheat temperature of steam is possible up to temperature of gas mixture at gas turbine outlet with pinch temperature (T_{gto} - ΔT) as shown in Figure 3. At this condition, steam is supplied to combustion chamber at highest possible temperature, heated using gas turbine exhaust flue gas mixture, which results in S_{rmin} with maximum efficiency. This maximum possible temperature is limited by temperature of the flue gases and further heating of mixture of steam and flue gases is carried out in combustion chamber. On the other side, for saturated condition, steam heated from gas turbine exhaust flue gas mixture is at saturated condition supplied to combustion chamber for S_{rmax} for lower efficiency but higher net work. These conditions are represented in Figure 4. Region of S_r lower than S_{rmin} is showing high stack losses and region higher than S_{rmax} is presenting steam quality at the entry to combustion chamber as wet.

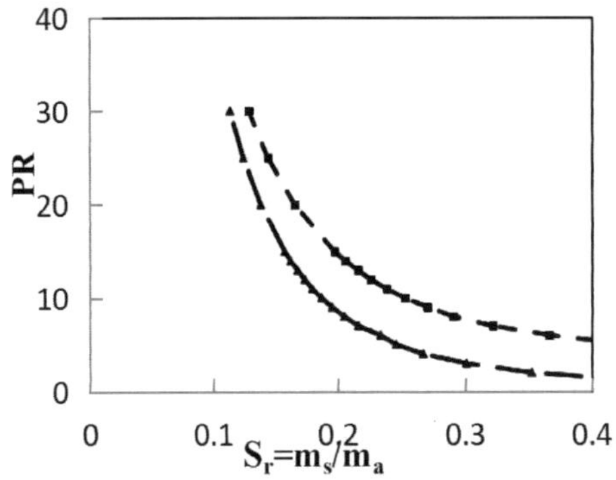

Figure 4. Integration at S_{rmax} and S_{rmin}

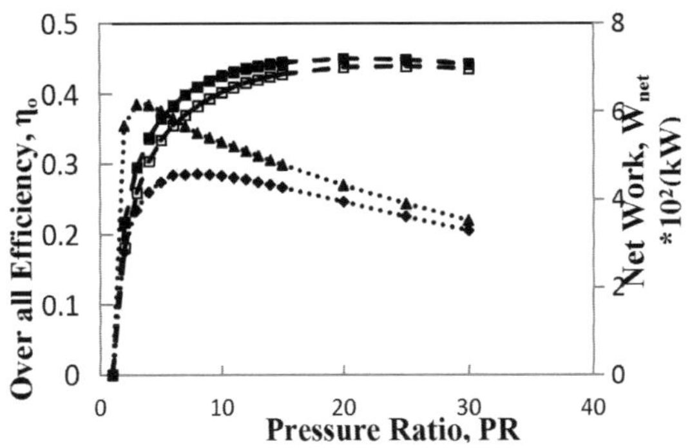

Figure 5. Efficiencies and net work at S_{rmax} and S_{rmin}

Figure 5 is useful for selecting the appropriate steam condition so that selected system is optimal for given condition. For S_{rmin}, with PR between 8-9, efficiency is optimal and W_{net} is maximum (at S_r =

0.205). However, for S_{rmax}, W_{net} is maximum at lower pressure ratio (PR=3), but with lower efficiency. For entire range of PR, efficiency is higher for minimum S_r, and W_{net} is higher for S_{rmax}.

6. Conclusions:

The pinch analysis based methodology is proposed for investigation on steam injected gas turbine (STIGT) cycle. The proposed methodology is very useful for appropriate selection of steam ratio required for design of STIGT. The new plot of PR verses steam ratio (S_r) gives various regions of integration showing the appropriate space for feasible integration where design parameters can be selected based on objectives. Additionally, plot of overall efficiency and net work verses steam ratio is utilized for new system design as well as retrofitted system. For retrofitted system, at given pressure ratio, there are two objectives either maximum efficiency or maximum net work for design of STIGT system. In case of new design, initially, objective is fixed and based on it PR can be determined. Practically, the limiting value of steam ratio is 0.3. Considering this limitation, appropriate pressure ratio can be selected. For maximum overall performance, it is advisable to go for a STIGT plant with S_{rmin} at optimal PR in the range of 7 to 9 for an illustrative problem. On the other side, for maximum steam ratio with same condition, W_{net} is slightly higher compared to S_{rmin}. Higher efficiency indicates less pollution due to flue gases leads to make environment clean and have more possibility to adhere the stringent environmental norms. Higher net work output indicates more energy available for use but with little lower efficiency.

References

[1] Najjar Y S H 2000 *Appl. Therm. Eng.* **20(2)** pp 179–197
[2] Heppenstall T 1998 *Appl. Therm. Eng.* **18(9-10)** pp 837–846
[3] Poullikkas A 2005 *Renew. Sustain. Energy Rev.* **9(5)** pp 409–443
[4] Kenney W F 1984 *Energy conservation in the process industries Academic Press*
[5] Horlock J H 1987 *Combined Heat and Power (Thermodynamics and fluid mechanics series) Pergamon Press Oxford*
[6] Jarre M, Noussan M and Poggio A 2016 *Appl. Therm. Eng.* **100** pp 304–314
[7] Alqahtani B J and Patiño-Echeverri D 2016 *Appl. Energy* 169 pp 927–936
[8] Mohammadi Khoshkar Vandani, A Joda F and Bozorgmehry Boozarjomehry R 2016 *Energy Convers. Manag.* **109** pp 103–112
[9] Sahin A Z , Al-Sharafi A, Yilbas B S and Khaliq A 2016 *Energy Convers. Manag.* **116** pp 91–100
[10] Ali U , Agbonghae E O, Hughes K J, Ingham D B , Ma L and Pourkashanian M 2016 *Appl. Therm. Eng.* **103** pp 747–58
[11] Olsommer B, Favrat D and Spakovsky M R Von 1999 *Int.J. Appl. Thermodyn.* **2(3)** pp 97–114
[12] Srinivas T, Reddy B V and Gupta A V S S K S 2011 *Int. J. Thermodyn.* **14(1)** pp 29–36
[13] Alves L G and Nebra S A 2003 *Int.J. Thermodyn.* **6(1)** pp 13–22
[14] Burnham J B, Giuliani M H and Moeller D J 1987 *J. Eng. Gas Turbines Power-Transactions ASME* **109(3)** pp 257–62
[15] Foster-pegg 1989 *The Gas Turbine and Aeroengine Congress and Exposition ASME Toronto Ontario Canada*
[16] Bolland O and Stadaas J F 1995 *J. Eng. Gas Turbines Power-Transactions ASME* **117(1)** pp 138–45
[17] Horlock J H 2003 *Advanced Gas Turbine Cycles Oxford*
[18] Aichmayer L, Spelling J and Laumert B 2015 *Energy Procedia* **69** pp 1089–99
[19] Rio M S del, Lucquiaud M and Gibbins J 2014 *Energy Procedia* **63** pp 2530–41
[20] Canepa R and Wang M 2015 *Appl. Therm. Eng.* 74 pp 10–19
[21] Bade M H and Bandyopadhyay S 2015 *Appl. Therm. Eng.* **78** pp 118–28
[22] Fagbenle R L, Oguaka A B C and Olakoyejo O T 2007 *Appl. Therm. Eng.* **27(13)** pp 2220–25

IOP Conf. Series: Materials Science and Engineering **282** (2017) 012009 doi:10.1088/1757-899X/282/1/012009

DECO FRECASE (drywall eco-friendly from eggshell and cane bagasse) as an innovation of eco-friendly interior construction

M A Imron[1], D N I Ahkam[2] and A W Hidayat[2]

[1]Departement of Economics, Islamic University of Indonesia, Sleman, Yogyakarta 55283, Indonesia.
[2]Departement of Civil Engineering and Planning, Islamic University of Indonesia, Sleman, Yogyakarta 55584, Indonesia

E-mail: aliimron@students.uii.ac.id

Abstract. The number of factories and home industries, both upper and lower middle class certainly adds waste generated, resulting in environmental pollution. The development of buildings is one of the largest contributors to global warming. For that, it takes technological innovations that lead to the criteria of green building. The application of green material is important aspects of environmentally friendly development, the selection of materials on the green material criteria of both roles should be applied continuously in order to realize the environmental sustainability of the material. Utilization Waste eggshell and bagasse which is a community waste, has the potential to become innovative environmentally friendly building materials. The eggshell is composed of 94% calcium carbonate, 1% magnesium carbonate, 1% calcium phosphate, and 4% organic material, especially protein, while the bagasse has a high content of silica (SiO2). In this study, the compounds are used as raw material for making alternative drywall in the form of DECO FRECASE. DECO FRECASE is an innovation of environmentally friendly building materials as an interior wall construction. Through DECO FRECASE, it is expected that building material innovation in Indonesia can be improved and of course environmental problems can be minimized by utilizing it as raw material for building construction.

1. Introduction

The national development of Indonesia aims to create a just and prosperous society. Implementation needs to consider the balance of the three pillars of sustainable development, namely social, economic and environmental. With so the development will be in accordance with the rules of the environment, namely increasing the value and function of the environment. The implementation of environmental management involves all stakeholders, so it needs the carrying capacity, capacity and reserve aspects and spatial so as not to cause various environmental disasters such as environmental pollution. The number of factories and home industries, both upper and lower middle class certainly adds waste or waste generated, resulting in environmental pollution. For example, the expansion of sugarcane production in the state of São Paulo has been shown to cause an increase in the local abundance of rodents and result in the emergence of infectious diseases such as hantaviruses and leptospirosis in human[1]. The large number of food companies and sugar mills in Yogyakarta produces an abundance of waste that pollutes the environment and in the absence of innovations to utilize waste, So that people feel uncomfortable with the surrounding environment and concern for the environment is declining. Therefore, we take the initiative to innovate from this waste into valuable and environmentally friendly materials. Maintenance of the environment must be comprehensive, such as in the process of infrastructure development. The development of buildings is one of the largest contributors to global warming. For that, it takes technological innovations that lead to the criteria of green building. The application of green material is important aspects of environmentally friendly development, the selection of materials on the green material criteria of both roles should be applied continuously in order to realize the environmental sustainability of the material. Utilization Waste egg shells and bagasse is a community waste that has the potential to be innovative building materials that are environmentally friendly.

1.1 Drywall

Drywall is an indoor partition or wall system consisting of a gypsum board mounted on a frame with the help of special couplers. Its installation does not use a mixture of cement and air, hence it is called drywall. Very different from conventional walls in general that use bricks, mixture of cement, sand, and air.

2. Problems

1. How does the DECO FRECASE concept as an innovation of waste pollute the environment into eco-friendly interior construction?
2. How SWOT analysis of DECO FRECASE as innovation of environmentally friendly interior construction?

3. Objective

The main objective of this research is to innovate waste that pollute the environment into a valuable and environmentally friendly building material through the creation of DECO FRECASE.

4. Material

This research was to combine eggshell and bagasse as an environmentally friendly building material and reduce environmental pollution.

4.1 Eggshell

The eggshell structure consists of three layers, namely the cuticle layer, sponge layer, and lamellar layer. The cuticle layer is the outermost surface containing a number of proteins. The layers of sponges and lamellar form a matrix formed by protein fibers bound by calcium carbonate in an eggshell. The eggshell represents 11% of the total egg weight and is composed by calcium carbonate (94%), calcium phosphate (1%), organic material (4%), and magnesium carbonate (1%)[2].

The content of calcium carbonate from dried chicken eggshells with albumin paste can be seen in the table below.

Table 1. Nutrient composition of dried chicken eggshells with albumin paste [3].

Nutrition	Content (% weight)
Water	29-35
Protein	1,4-4
Calcium	35,1-36,4
CaCO	90,9
Magnesium	0,37-0,40
Potassium	0,10-0,13
Sulfur	0,09-0,19

4.2 Bagasse

Sugarcane is one of the silicone collecting plants (Si) which is its Si uptake plant exceeds its absorption to water. During the growth (1 year), the cane absorbs Si about 500-700 kg per ha higher than other elements. For comparison, within the same time the cane absorbed between 100-300 kg K, 40-80 kg P, and 50-500 kg N per ha[4].

The forming structure of bagasse fiber consists of Cellulose, Hemicellulose, Pentosans and Lignin which composition as follows

Table 2. Structure of bagasse ((Lacey,J. *The Microbicloby of the Bagasse of Sugar Cane-Proc.of XVII Congress of ISSCT*)

No	Component	Percentage of Dry Weight
1	Cellulose	26%-43%
2	Hemicellulose	17%-23%
3	Pentosans	20%-33%
4	Lignin	13%-22%

Looking at the composition of bagasse at table 2, the bagasse fiber has the most cellulose content and cellulose is a sugar-containing content.

5. Method

Our research held on Special Region of Yogyakarta which takes place within 3 weeks. Starting from the first week until the third week of May 2017. Our research stages are started from research planning, research preparation, material collection, product making, drying product, product testing, writing preparation. The materials used in this study are eggshell, ashdod bagasse, dry sugar cane fiber, white portland cement, water, and additional adhesives (optional). Variables in this study are variation of egg shell addition as gypsum substitute and variation of addition of bagasse material. While the test parameters include density, water uptake, and compressive strength.

Preparation The samples used in the manufacture of DECO FRECASE consist of egg shell, bagasse ash, dried sugar cane and cement fiber. To know the composition of raw materials refers to the combination of the proportion of gypsum board (drywall) in general with partition walls. So to facilitate the mixing process then all the composition of the raw material is determined in the percentage of volume. In this study, the matrix used was eggshell powder, cement adhesive mixture, and bagasse ash while aggregate consisted of dried sugarcane fiber as a replacement material around.

Table 3. Sample Composition of Sample Preparation Sample[3]

[3]For egg shells and percentage of bagasse ash from matrix volume, while for dry sugarcane fiber and cement percentage of drywall volume

Sample		**Percent Composition**		
	Eggshell	Bagasse Ash of Sugar Cane	Cane Fibers	Cement
A	25%	75%	0% - 30%	10% - 30%
B	50%	50%	0% - 30%	10% - 30%
C	75%	25%	0% - 30%	10% - 30%

The DECO FRECASE research stage starts from the shell collection and egg shell. Egg shell is a production raw material that can affect the strength of DECO FRECASE. Therefore, the collection of large amounts of eggshell waste is needed in the manufacture of DECO FRECASE. To collect this eggshell, we get from bakery which is base of industry in Yogyakarta. The eggshell is then washed clean to avoid the stench and bacteria in DECO FRECASE. The purpose of this shell's shell is to facilitate the process of smoothing the egg shell into a powder that is completely dry and non-sticky.

Furthermore collection and burning of sugarcane. Sugarcane luggage is partial replacement of White Portland cement in our products. In addition, bagasse makes heavy DECO FRECASE lighter so it can be used for earthquake prone areas. Therefore, the collection of bagasse in large quantities greatly affects the weight of DECO FRECASE. We get this bagasse from sugar cane factory waste in Yogyakarta which the waste is left alone, thus polluting the environment. Bagasse is collected, dry for 7 days. The dry sugarcane bagasse is for the fulfillment of the needs of bagasse ash and dried sugarcane fiber. After bagasse pulp is dried, bagasse is processed into 2 stages. Some of the combustion process is done while the other half is separated between the sugarcane fibers with the skin to be taken his fiber as a replacement material.

The combustion process begins with burnt bagasse or forming coal by placing enough dregs of waste into the fuel barrel. Once the embers are formed, add little by little dried bagasse to the barrel by pushing it with a stick so it is rather dense. Burning is done for 4-5 hours so that bagasse becomes ash. The next step is to wait for the cold ash and then filter the ashes to separate ash and other solid particles to get the same ash and particle size. Next is egg shell refinement. Egg shell refinement aims to get a less rough DECO FRECASE texture. Egg shell refinement is done by colliding the eggshell in the form of large fractions to a fine powder. After being smoothed, sieving with a tea filter to separate fine powder with a coarse shrapage so that the fraction can be roughened again.

DECO FRECASE dough making of each raw material is measured according to the predetermined composition. Once tested, bagasse ash, eggshell powder, dried sugarcane fiber, and cement are mixed in a container and stirred evenly using a cement spoon or mixer. Then add water until the mixture is mixed. The good dough texture in the making of DECO FRECASE is not very fluid and sticky. It is intended that the drying stage is not too long because the evaporation of water that is not too much and facilitate the printing.

In the manufacture of DECO FRECASE dough using adhesive in the form of white portland cement because the material is more affordable. Although adhesives continue to use cement white portland, the use of bagasse as a substitute for portland cement as the main ingredient can reduce portland cement use by up to 50%.

In DECO FRECASE printing stage, the DECO FRECASE dough is applied to the block mold. Prior to printing, the mold is first smeared with oil to facilitate the release of the dough from the mold after the drying step. The sample test mold used in this research is a 6 x 6 x 3 cm beam aluminum foil mold. First, pour the dough until it fills half the mold and flatten the DECO FRECASE dough. After that, pressed to the dough to make the dough more dense and not easy to crack. Printing of DECO FRECASE should really be noted because printing affects the durability of DECO FRECASE.

Next stage Drying DECO FRECASE. The treatment of the printed drywall was carried out by the drying process naturally at room temperature for 7 days, then the test sample was dried in the oven kitchen at 250°C with a holding for 1 hour. The DECO FRECASE drying is done to obtain a strong DECO FRECASE texture so that even if it has light weight it can be used as a partition wall.

Last Step is DECO FRECASE Testing. After the specimen is subjected to hardening process, then the test is done. To know the characteristics of the concrete, it is necessary to measure or test the physical and mechanical quantities. Testing of DECO FRECASE characteristic in this research include: density, water absorption, and compressive strength. The density measurements of each of the prepared DECO FRECASE sample compositions, observed using the object type mass principle.

6. Discussion

6.1. Drywall Eco-Friendly From Eggshell and Cane Bagasse (DECO FRECASE)

Drywall Eco-Friendly From Eggshell and Cane Bagasse (DECO FRECASE*)* is an alternative eco-friendly drywall product made from raw eggshells integrated with bagasse as an interior wall construction. The alternative understanding in DECO FRECASE is to innovate environmentally friendly building materials and reduce environmental pollution by reducing organic waste.

The test results prove that the DECO FRECASE product in the water-absorption test is proven that DECO FRECASE products tend to increase with the increase of eggshell composition and bagasse ash. In contrast, the compressive strength test and density test tend to decrease as the eggshell composition and bagasse ash are added. Therefore, proper composition is required for DEFFEC density to reach drywall standard with composition according to sample A.

6.2. SWOT Analysis

Based on the SWOT analysis approach, DECO FRECASE has an internal analysis including an assessment of the strength factor (Strength) and weakness (Weakness). Meanwhile, external analysis includes opportunity (Opportunity) and challenge (Threaths). Here's the assessment:

Table 4. *SWOT* analysis approach based on Strength and Weakness

No.	*Strength*	*Weakness*
1.	Production raw materials are easy to obtain because they come from wastes that are rampant in the community so have a relatively cheap price	Long drying due to the strong ash bagasse binds water
2.	More environmentally friendly because it utilizes the existing waste.	
3.	Easily manufactured without the need for special skills..	
3.	Applicative as partition wall.	
4.	Great market opportunity because of the construction material in the form of drywall is one of the products needed by the community.	
5.	Can reduce the use of bricks and portland cement in the main building materials..	
6.	Lighter than the drywall in general so it can be used in earthquake prone areas.	

Solutions to take are work with drywall producers, architects, and experts in determining suitable materials to accelerate drying at DECO FRECASE and conducting further research on appropriate techniques in drying DECO FRECASE so that obtained the process of rapid drying and target results.

Table 5. *SWOT* Analysis approach based on Opportunity and Threats

No	*Opportunity*	*Threaths*
1.	Being a product of innovation technology that is in demand by the community because along with the growth of the population the need for interior wall construction is increasing as it is practical and efficient	The DECO FRECASE is in the marketing stage.
2.	Optimizing the productivity of society because it can be used as an opportunity for entrepreneurship and open employment.	
3.	Being innovated building materials that utilize waste but according to standards.	
4.	Reduce waste problems in the environment due to the efforts to utilize waste egg shells and bagasse into a more economical product.	

Solutions to take are governments and research institutes and academics are seeking the introduction and promotion of DECO FRECASE on a wider scale to better recognize the wider community and cooperate with bakeries and communities to collect eggshells and bagasse so as to obtain the optimal waste of eggshells and bagasse every day.

7. Conclusion

From the problems and discussions that have been described above, it can be concluded that Drywall Eco-Friendly from Eggshell and Cane Bagasse (DECO FRECASE) is an alternative eco-friendly drywall product made from raw eggshells integrated with bagasse as an interior wall construction that has the potential to become an economic product that can improve the innovation technology. DECO

FRECASE also through the density, water absorption, and compressive test proves that DECO FRECASE with composition according to sample A meets the drywall standard.DECO FRECASE has many advantages, such as drywall designed more environmentally friendly because it utilizes existing waste, can reduce the use of bricks and portland cement in the main building materials, easy to apply in the manufacture of building barriers, easy to produce without the need for special skills, and Production raw materials are easy to obtain.

References

[1] Luciano M. Verdade, Carla Gheler-Costa, Marli Penteado and Graziela Dotta. 2012. The Impacts of Sugarcane Expansion on Wildlife in the State of São Paulo, Brazil (*Journal of Sustainable Bioenergy Systems* vol **2**) PP 138-144. http://dx.doi.org/10.4236/jsbs.2012.24020

[2] Rivera, E M , *et al.* 1999. *Sythesis of hydroxyapatite from eggshells.* Meksiko: Uniersidad Nacional Autonoma de Mexico

[3] Ockerman, H W dan Hansen and C L 2007. *Nutrient Composition of Dried Eggshells (with Adhering Albumin).* Bioinfobank Library 200:7

[4] Edo Yukamgo and Nasih Widya Yuwono 2007 Peran Silikon Sebagai Unsur Bermanfaat pada Tanaman Tebu. (*Jurnal Ilmu Tanah dan Lingkungan vol* **7**) pp 103-116

A Study of Commuters' Decision-Making When Delaying Departure for Work-Home Trips

Fangjie Que and Wei Wang

School of Transportation, Southeast University, Nanjing 210096, China

E-mail:wangwei_transtar@163.com

Abstract. Studies on the travel behaviors and patterns of residents are important to the arrangement of urban layouts and urban traffic planning. However, research on the characteristics of the decision-making behavior regarding departure time is not fully expanded yet. In this paper, the research focuses on commuters' decision-making behavior regarding departure delay. According to the 2013 travel survey data of Suzhou City, a nested logit (NL) model was built to represent the probabilities of individual choices. Parameter calibration was conducted, so that the significant factors influencing the departure delay were obtained. Ultimately, the results of the NL model indicated that it performed better and with higher precision, compared to the traditional multinomial logit (MNL) model.

1. Introduction

Departure time prediction is an important component when studying commuters' travel decision-making behavior. Facing serious traffic congestion during daily peak periods, more people choose later departure time after work or school. Congestion charging, flexible schedules, and other traffic-demand management strategies were adopted in several cities, which provide commuters a wider range of choices in departure time and a greater utility of delaying departure. A discrete choice model was created in this paper, to help analyze the characteristics of resident travel behaviors and master the distribution regulation of departure time. The results can be used to predict travel behaviors and evaluate the effect of urban planning and management strategies.

2. Nested-Logit model construction method

The nested logit (NL) model, first derived by Moshe Benakiva [1], is a discrete choice model designed to capture correlation among alternatives. Unlike the traditional multinomial logit (MNL) model, it divides choice sets into nests, and the alternatives within each nest are correlated [2]. For example, in a two-level NL model, alternatives D are divided into nests M, based on the assumption that the alternatives in the same nest share the same variance of deviation [3]. Logically, individual n consider of choosing nest m, $m \in M$, before choosing the specific alternative d, $d \in D$, as the choice process.

Let the utility of the alternative dm chosen by commuter n be equation (1),

$$U_{dmn} = V_{dmn} + V_{mn} + V_{dn} + \varepsilon_{dmn} + \varepsilon_{mn} \tag{1}$$

where the deterministic components V_{dmn}, V_{mn}, and V_{dn} are defined as a linear combination of the measurable variables $\boldsymbol{X}_{dm}$, $\boldsymbol{X}_m$, and $\boldsymbol{X}_d$, with undetermined coefficients β_{dm}, β_m, and β_d, respectively. ε_{dmn} and ε_{mn} are random utilities.

It is assumed that ε_{dmn} follows a Gumbel distribution with scale parameter μ_d, while $\max_{d\in D} U_{dm}$ follows a Gumbel distribution with scale parameter μ_m. All people are assumed to choose the alternative providing the greatest utility.

Following the choice process mentioned above, the probability of the alternative *dm* chosen by commuter *n* is defined as a conditional probability shown as equation (2),

$$P_n(dm) = P_n(d \mid m) P_n(m) \tag{2}$$

The probability of the alternative *d* if nest *m* is chosen is expressed as equation (3),

$$P_n(d \mid m) = \frac{e^{(V_{dmn}+V_{dn})\mu_d}}{\sum_{d' \in D_{rn}} e^{(V_{d'mn}+V_{d'n})\mu_d}} \tag{3}$$

And the probability of nest *m* is **e**quation (4),

$$P_n(m) = \frac{e^{(V_{mn}+V_{mn}^{*})\mu_m}}{\sum_{m' \in M_n} e^{(V_{m'n}+V_{m'n}^{*})\mu_m}} \tag{4}$$

where V_{mn}^{*} is the logsum in equation (5),

$$V_{mn}^{*} = \frac{1}{\mu_d} ln \sum_{d \in D_{rn}} e^{(V_{dmn}+V_{dn})\mu_d} \tag{5}$$

These functions can be integrated into one equation and rewritten as equation (6),

$$P_n(dm) = P_n(d \mid m) P_n(m) = \frac{e^{(V_{dmn}+V_{dn})\mu_d}}{\sum_{d' \in D_{rn}} e^{(V_{d'mn}+V_{d'n})\mu_d}} \frac{e^{(V_{mn}+V_{mn}^{*})\mu_m}}{\sum_{m' \in M_n} e^{(V_{m'n}+V_{m'n}^{*})\mu_m}} \tag{6}$$

3. Departure delay choice modeling

3.1. Building selection tree

According to the commuter travel survey in the Central Business District of Suzhou, Jiangsu, the distribution of departure delay is shown in Figure 1.

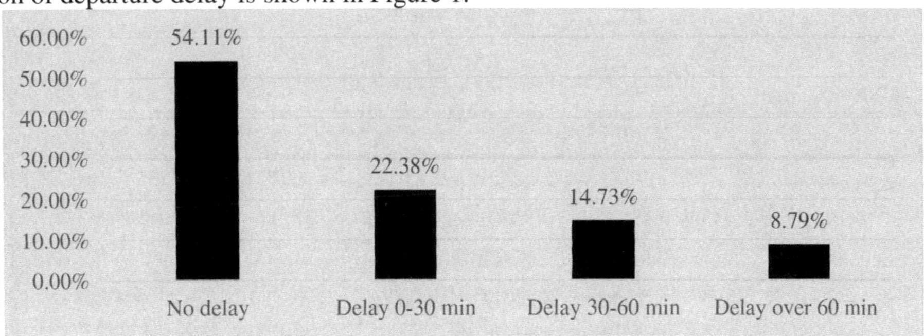

Figure 1. Histogram of the departure delay for work-home trips.

As can be seen from Figure 1, there are four available alternatives for departure delays for work-home trips. Nearly half of commuters delay departure for work-home trips. People who choose delay departures within 30 minutes comprise 22.38% of the population, while 8.79% of commuters delay departures greater than one hour. The alternatives can be partitioned into two nests *M*, *M*= {delay, no delay}. The former nest only includes one alternative, while the latter comprises three choices: delay departure within 30 minutes, 30-60 minutes, and over one hour, numbered 12, 22, and 32, separately. The nested structure is shown in Table 1.

Table 1. The nested structure of the departure delay choice model.

Upper Level	Lower Level	Number	Percentage
No delay(1)	No delay (11)	665	54.11%
delay(2)	Delay 0-30 min(12)	275	22.38%
	Delay 30-60 min(22)	181	14.73%
	Delay over 60 min (32)	108	8.79%
Sum		1229	100%

3.2. Selecting variables

In this paper, nine factors were chosen as explanatory variables, after analysis of the survey data. The definitions and explanations of variables are listed in Table 2.

Table 2. Explanatory variables of the NL model.

Attribute	Variable	Variable Explanation
Individual	*gender*	1-male, 0-female
	age	Age
	income	Income
	ftime	If flexible work hours: 1-yes, 0-no
Household	*car*	Number of cars in household
	child	Number of children in household
Trip	*distance*	Distance from work to home (in miles)
	mode	Mode of transportation used work-to-home: 1-car SOV (single occupancy vehicle), 2-carpool, 3- vanpool, 4-bus,5-other.
Alternative-specific	*tatf*	Ratio of actual travel time to free-flow travel time

3.3. Parameter estimation

Data analysis and parameter estimation were completed using the Stata 12.0 application. With the base alternative of no delay, the final results are shown in Table 3.

Table 3. Final results of parameter estimation.

Lower level	Delay 0-30 min		Delay 30-60 min		Delay over 60 min	
constant	-7.77	(-3.53)	-11.95	(-3.90)	-13.33	(-5.53)
tatf	-13.05	(-7.02)	-13.05	(-7.02)	-13.05	(-7.02)
gender	—	—	—	—	-2.41	(-1.88)
age	0.06	(1.88)	-0.08	(-1.72)	—	—
ftime	—	—	2.29	(2.17)	—	—
distance	-0.35	(-2.59)	—	—	—	—
mode1	2.75	(2.40)	4.29	(2.60)	—	—
mode2	—	—	3.60	(1.78)	—	—
mode3	—	—	—	—	3.88	(2.19)
Upper level			Delay			

child	0.19	(2.78)
distance	0.28	(3.43)
mode4	2.17	(2.11)
Dissimilarity parameters:		
$\tau_{nodelay}$	1	—
τ_{delay}	0.67	4.45***
LR test for IIA (tau = 1):	*chi2(2) =9.79*	*Prob > chi2=0.007 5*

Near the end of the output, dissimilarity parameters exist, which measure the degree of correlation of random shocks within each of the two alternatives in the upper level. A likelihood-ratio test, with the null hypothesis that all parameters are one, can also be determined. Equivalently, the property known as the IIA [4], imposed by the multinomial logit model holds if and only if all dissimilarity parameters are equal to one [5]. The dissimilarity parameter in this model is 0.67, passing the significance test. Therefore, the null hypothesis is rejected, indicating that the alternatives in nest "delay" are not independent.

The equations of utility and *logsum* in the lower level can be determined by the results of the estimation in Table 3.

Utility values of alternatives in the lower level are given by:

V_{12}=-7.77-13.05*tatf*+0.06*age*-0.35*distance*-2.75*mode1*,

V_{22}=-11.95-13.05*tatf*-0.08*age*+2.29*ftime*+4.29*mode1*+3.60*mode2*,

V_{32}=-13.33-13.05*tatf*-2.41*gender*+3.88*mode3*,

and the *logsum* can be calculated based on equation (5):

V_2^*=0.67ln[exp(-7.77-13.05*tatf*+0.06*age*-0.35*distance*-2.75*mode1*)
+exp(-11.95-13.05*tatf*-0.08*age*+2.29*ftime*+4.29*mode1*+3.60*mode2*)
+exp(-13.33-13.05*tatf*-2.41*gender*+3.88*mode3*)].

The relative utility function of delay departure is expressed as:

V_2=0.19*child*+0.23*distance*+2.17*mode4*.

Ultimately, the probabilities of departure delay alternatives were computed by the expressions in Table 4.

Table 4. Probability expressions of each departure delay alternative for work-home trips.

Upper Level	Lower Level	Probability
No delay (1)	No delay (11)	$\frac{1}{1+\exp(-V_2-V_2^*)}$
	Delay 0-30 min (12)	$\frac{1}{1+\exp(V_2+V_2^*)}\cdot\frac{\exp(V_{12})}{\exp(V_{12})+\exp(V_{22})+\exp(V_{32})}$
Delay (2)	Delay 30-60 min (22)	$\frac{1}{1+\exp(V_2+V_2^*)}\cdot\frac{\exp(V_{22})}{\exp(V_{12})+\exp(V_{22})+\exp(V_{32})}$
	Delay over 60 min (32)	$\frac{1}{1+\exp(V_2+V_2^*)}\cdot\frac{\exp(V_{32})}{\exp(V_{12})+\exp(V_{22})+\exp(V_{32})}$
	Sum	1

4. Discussion

4.1. Bottom model parameters

Table 3 shows that there are eight factors significantly influencing the decision-making in the lower level.

Considering the variables with individual attributes, gender only affects the choice of delaying departure for more than one hour and, on equal conditions, women are more likely to delay departure for more than 60 minutes than men. The estimates of age were 0.062, -0.079, -, indicating that the elderly prefer not to go home too late. Flexible schedules provide more choices for departure time, and the utility of choosing a 30-60 minutes' delay significantly increases.

The family characteristics variables, including number of children and cars in household, have little effect on the length of delay.

Commuter distance influences when people choose a departure delay of less than 30 minutes. Regarding the mode of transportation used from work-to-home, car SOV, carpool, and vanpool affect the alternative No. 12, 22, and 32, respectively, if the alternative of delay were chosen. Consequently, reasonable resource allocation of both time and space must be provided when developing and promoting ride sharing.

Of all the factors in the lower model, the most significant factor affecting the departure time is the ratio of the actual travel time to free-flow travel time, which is a determinant for commuters when making decisions on the departure time for work-home trips.

4.2. Upper model parameters

The number of children in the household, distance from work to home, and use of buses are the main factors in the upper model. Commuters with more children are more likely to depart on time. People with shorter commute distances tend to delay departure. Therefore, developing new towns and constructing polycentric urban structures allow people to choose a wider variety of work places and make diversified schedules. For commuters who choose buses as their transport, the utility value of delay will be greatly reduced. Hence, a reasonable arrangement of bus schedules is important to commuters' decision-making regarding departure time.

4.3. Prediction accuracy

To evaluate the model, the accuracy of model prediction can be described by two indexes: sensitivity and specificity. The receiver operating characteristic curve (ROC) is a comprehensive evaluation of the sensitivity and specificity of the established model, utilizing geometric methods to reveal the relationship between the two indexes [6]. Generally, the value of the area under the curve (AUC) is between 0.5 and 1. The closer the area is to 1, the higher the prediction accuracy of the model. In this paper, the ROC curve of the NL model is shown in Figure 2.

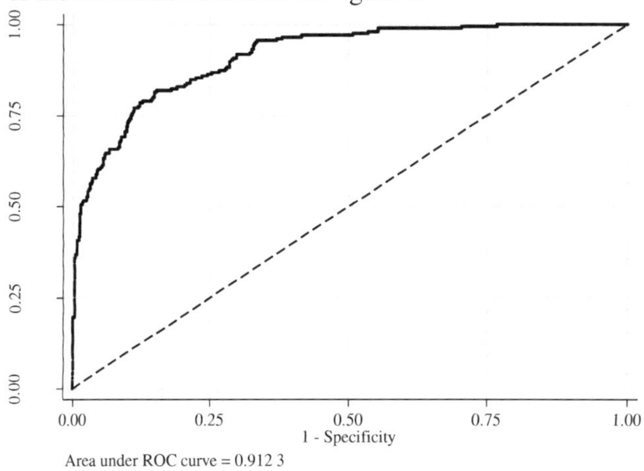

Figure 2. ROC curve.

The area under the ROC curve (AUC) is 0.9123, which indicates the model prediction has high accuracy.

5. Conclusion

The principle and formula derivation of the NL model is introduced in this paper, and is applied to the study of decision-making regarding departure delays. After analysis of the characteristics of commuters' delays on work-home trips, the structure of the delay alternative tree is provided. According to survey data in Suzhou City, the parameter calibration is conducted via Stata. The results show that the model has high accuracy, and describes the characteristics of commuters' decision-making regarding delaying departure on work-home trips better than the traditional MNL model, overcoming the defects of the independence of irrelevant alternatives, which has a reference and adaptive significance.

To acknowledge the travel time characteristics more comprehensively and scientifically, and understand more practical approaches to solving decision-making behavior problems, further research could be performed in the following respects. First, the quantity of explanatory variables should be enriched to improve the model, adding cost and level of service of roads as independent variables. Next, the model can be applied to the evaluation of traffic management measurements, such as the before-after study of congestion pricing. Also, the model could be used for urban traffic demand forecasting and transportation planning evaluation, which has distinctly practical value.

References

[1] Moshe B 1974 *J.* Structure of passenger travel demand models. *Transportation Research Board Record.* **526** pp 26-42
[2] Daniel M 1974 Conditional logit analysis of qualitative chioce behavior *Frontiers in Econometrics* ed Zarembka P (New York: Academic Press) pp 105–142
[3] Moshe B and Steven R L 1985 *Discrete choice analysis: theory and application to travel demand* (London: MIT Press)
[4] Hensher D and Greene W 2002 *J.* Specification and estimation of the nested logit model: alternative normalizations. *Transportation Research Part B: Methodological.* **36(1)** 1-17
[5] Tuansheng C 2007 Travel behavior characteristics analysis for commuter. Doctoral Thesis. (Beijing: Beijing Jiaotong University)
[6] Mario A C 1999. Receiver operating characteristic (ROC) analysis. *Stata Technical Bulletin* **52** pp19-33

Destructive behavior of iron oxide in projectile impact

Wang Shang[1], Wang Xiaochen[1,*] ,Yang Quan[1] and Shan Zhongde[2]

[1]School of Engineering Technology, University of Science and Technology Beijing, Beijing 100083, China

[2]State Key Laboratory of Advanced Forming Technology and Equipment, China Academy of Machinery Science and Technology, Beijing 100083, China

*Corresponding author : wangxiaochen@ustb.edu.cn

Abstract. The damage strain values of Q235-A surface oxide scale were obtained by scanning electron microscopy (SEM/EDS) and universal tensile testing machine. The finite element simulation was carried out to study the destruction effects of oxidation at different impact rates. The results show that the damage value of the oxide strain is 0.08%. With the increase of the projectile velocity, the damage area of the oxide scale is increased, and the damage area is composed of the direct destruction area and the indirect failure area. The indirect damage area is caused by the stress/strain to the surrounding expansion after the impact of the steel body.

1. Introduction

In the hot rolling and subsequent cooling process, the surface of strip steel will form a dense, brittle oxide scale (commonly known as scales). Before the cold rolling production, the strip steel should pass acid washing process to remove the surface oxide scale and ensure that the finished product quality[1-2]. As the acidity general have the problems of less-pickling, over-pickling and etc, the discharge of waste liquid causes serious pollution to the environment, which is not the intention of the green manufacturing[3-4].

Projectile impact descaling is a typical acid-free descaling mode, and its operation mechanism is from the difference of the iron oxide and iron Body material properties, which is, brittle iron oxide sheet will fractures in the impact of direct cracking, while iron Body will produce a certain elastic-plastic deformation[5-7]. Iron oxide skin damage strain is an important parameter of the projectile impact decaling process, and scholars have done a lot of research in this area. Evans and Robertson[8-9] argue that the evolution of the internal defects of the oxide layer is the root cause of the damage, and based on the fracture mechanics theory they proposed iron oxide skin failure strain calculation formula. Nagl[10] experimentally analyzed the relationship between the fracture behavior of the oxide layer and the thickness of the oxide scale, and found that the failure stress/strain decreases with the increase of the thickness, and the larger the thickness of the oxide layer is, the larger the defect length is, and the experimental results Evans and Robertson[8-9] theoretical derivation is consistent.

On the production line as a large number of projectile impact at the same time high-speed, and damage the strip surface, while the oxide layer is very thin, and it is difficult to capture the behavior of the bullet impacting the scales. The finite element software is used to simulate the impact of single shot pills and some experience results are received. However, the model is generally very simple. For example, the finite element model proposed by the scholar Gao Zhi[11] does not set the oxide layer, only analyzed the impact of projectile towards steel, failing to get scaling damage principles; in the finite element model proposed by the scholar Ren Yujun[12] the scale is too thick(0.5 mm), and the direction of the impact is vertical, which deviates from the actual field.

This paper mainly carried out the following three aspects of the work:

(1).The damage strain value of Q235-A surface oxide scale was obtained by universal tensile tester and scanning electron microscope (SEM/EDS).

(2).Based on the destructive strain value obtained in Step 1, the model of projectile impact failure was established. The damage behavior of strip scale was calculated numerically according to different impacting speed.

(3).The damage effect of the oxide scale on the surface of the strip was observed by using the shot blasting experiment and the 4XB-TV metallographic microscope.

These work provides fundamental support for the design and optimization of projectile descaling process parameters.

2. Stretch experiment

The experimental material is the Q235-A hot-rolled narrow strip, coil width is 200mm, and thickness is 3.5 mm. The following samples are taken from the central position of the original coil, the chemical composition shown in Table 1.

Table 1. The chemical composition of carbon steel

Element	C	Mn	Si	S	P
w/%	0.216	1.250	0.183	0.041	0.0398

(1).Sectional observation test method

The samples were prepared with the length and width of 10 mm. The surface of the sample was cleaned with ultrasonic cleaning instrument, washed with alcohol, dried and cross sectioned. The cross section was observed by ZEISS ULTRA 55 scanning electron microscope.

(2).Tensile test method

The tensile test specimen was prepared by line cutting. The length direction (drawing direction) of the sample was in the rolling direction. The specific shape and size were shown in Figure 1. A specific strain was applied to the specimen using a US MIT MTS universal tensile tester. The displacement was controlled to be 0.01 mm/min, held for 1 min after the preset strain was reached. The sample was removed and the surface was wiped with an alcohol cotton swab. The surface morphology was observed by a plate-type metallographic microscope. Strain rises from 0.04%, with the step size of 0.005%.

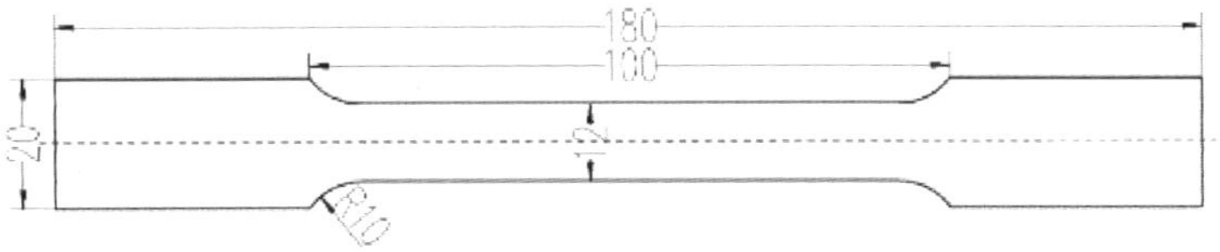

Figure 1. Tensile specimen dimensions (mm)

(3).Sample Cracking Observation Method

The tensile test specimens were measured to line cutting using an optical microscope to prepare a tensile test specimen with a length and a width of 10 mm. The surface of the sample was cleaned with an ultrasonic cleaner and washed with alcohol, and then dried. Then utilize a ZEISS ULTRA 55 scanning electron microscope to observe the surface of the sample.

3. Tensile Experiment and Discussion

3.1. Observation of cross section of iron oxide skin

Figure 2 shows the cross-sectional topography of Q235-A iron oxide under the scanning electron microscopy. The cross-sectional morphology of iron oxide skin shows: iron oxide skin has no obvious stratification, the thickness is relatively uniform, about 10 μm, the structure is relatively dense, and the Body is better[13].

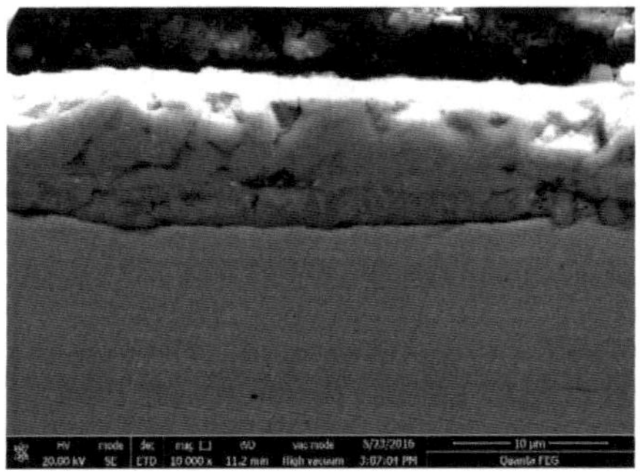

Figure 2. Cross-section morphologies of oxide scales

3.2. Tensile cracking behavior of iron oxide

The rupture behavior of iron oxide skin on the surface of each sample was observed by 4XB-TV metallographic microscope (200 times). When the strain was less than 0.08%, the surface oxide scale was not cracked, and the strain was obviously broken when the strain reached 0.08%.

When the specimen is cracked, multiple cracks occur simultaneously. Figure 3 shows a microscopic topography (10 000 times) of a 0.08% tensile strain observed by scanning electron microscopy. It can be seen from the micro-topography that the crack width is about 2.5 μm, the lines are neat and the debris that is peeled off during cracking. Obviously, the Q235-A iron oxide skin critical damage strain is 0.08%.

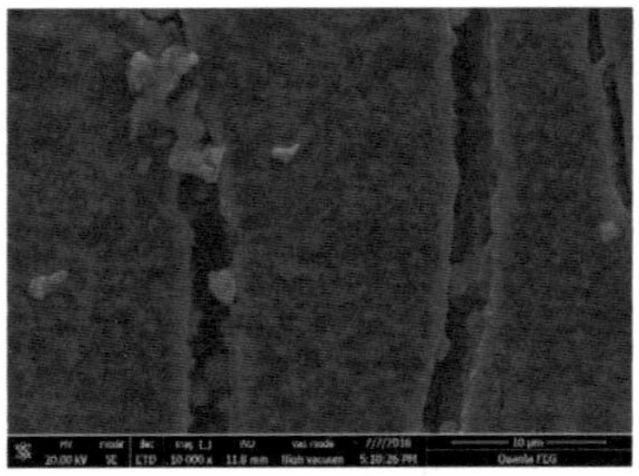

Figure 3. The crack of oxide scales

4. Finite Element Modeling

The 3D model of single projectile impact strip was established by ABAQUS/CAE finite element pretreatment software. The model has three parts: projectile, surface scale and strip. The radius of the projectile is 0.2 mm, the thickness of the surface is 0.02 mm, the radius is 0.5 mm, the thickness of the strip is 0.28 mm and the radius is 0.5 mm. Considering the symmetry of the whole structure, a 1/2 finite element model is established. In the unit style setting, the C3D10M type unit is applied for the projectile, the C3D8R type element is applied for the scale layer and the Body, and the TIE order[14] between the scale layer and the Body is bonded. After the grid is optimized, the finite element grid is shown in Figure 4.

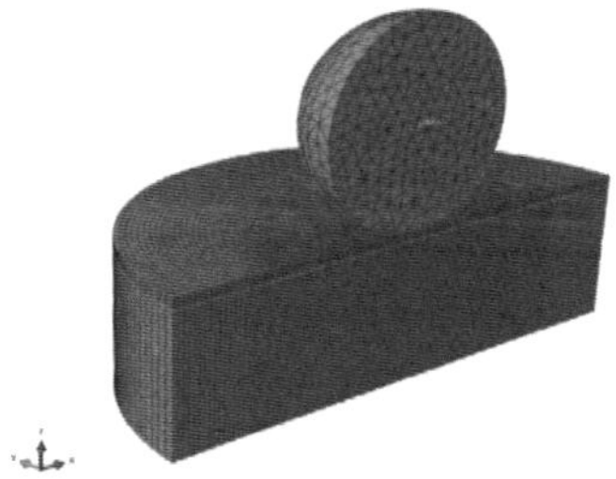

Figure 4. Schematic of finite grid model

In this model, the projectile is set as ideal rigid material, ignoring its irregular effects and deformation during striking. The scale is set to be brittle material without plasticity, and the failure criterion is ABAQUS brittle fracture failure[15]. The Body is set as an elastoplastic material, using the Johnson-Cook strength model[16]. The model is an empirical constitutive equation, and the Von Mises equivalent stress is a function of the equivalent plastic strain, the equivalent plastic strain rate and the temperature. The model equation[16] is shown in equation (1)

$$\bar{\sigma} = (\mathrm{A} + \mathrm{B}\bar{\varepsilon}^{pn})(1 + \mathrm{C}\ln\varepsilon^{*})(1 - \mathrm{T}^{*m}) \tag{1}$$

In this equivalent: A, B, n, C, m are material parameters with no units, and $\bar{\sigma}$ is equivalent stress; $\bar{\varepsilon}^{p}$ is equivalent plastic stress; ε^{*} is equivalent plastic strain rate with no units; T^{*} is temperature with no units, and its expression is

$$T^{*} = \frac{T - T_r}{T_m - T_r} \tag{2}$$

Here, T_r, T_m are the reference temperature and the melting point of the material respectively, T is current temperature.

The parameters of the model are shown in Table 2[17]. The simulation parameters and failure criteria are shown in Table 3[16,18].

Table 2. Parameters of Johnson-Cook models of Q235 steel.

A/Mpa	B/Mpa	n	C	m
244.8	899.7	0.94	0.0391	0

Table 3. Simulation parameters

	density / (kg·m^{-3})	Elastic Modulus /Gpa	Poisson's ratio	direct stress after cracking /Mpa	Direct stress damage strain /%
Projectile	7700	150	0.3	-	-
Body	7800	200	0.288	-	-
Scales	7750	210	0.29	180	0.08

The impact speed was set as 5 m/s, 10 m/s, 20 m/s, 30 m/s, 40 m/s and 50 m/s, respectively, with the impact angle (the angle between the projectile velocity direction and the strip surface for simulation) is 60°.

5. Introduction Finite Element Modeling

5.1. Effect Damage of Projectile Velocity on Scale

Figure 5 shows the impact of different projectile speed after the impact of strip surface damage layer cloud effect. In the calculation, the strain of the scale unit to break the strain is removed. Compared with the effect of scale damage at different speeds, the damage area increases with the increase of velocity. Under the impact of small speed, the model damage area is roughly semicircular, the damage area is very small. When the speed is 20 m/s, the semicircle of the damaged area begins to show "whisker-like crack" to the surrounding area, and the stress in the area of the "crack" is larger. When the velocity reaches 50 m/s, the larger scale is destroyed, and the "whisker" in the failure zone is close to the edge of the model.

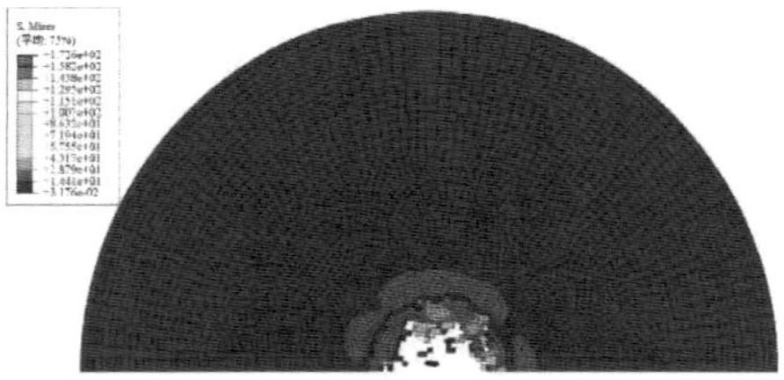

(a) v=5 m/s

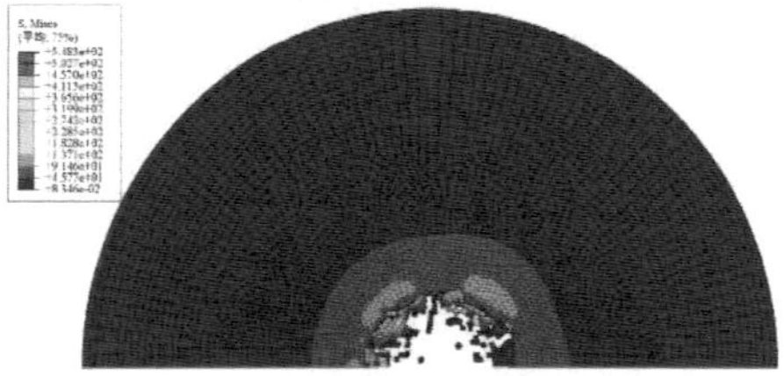

(b) v=10 m/s

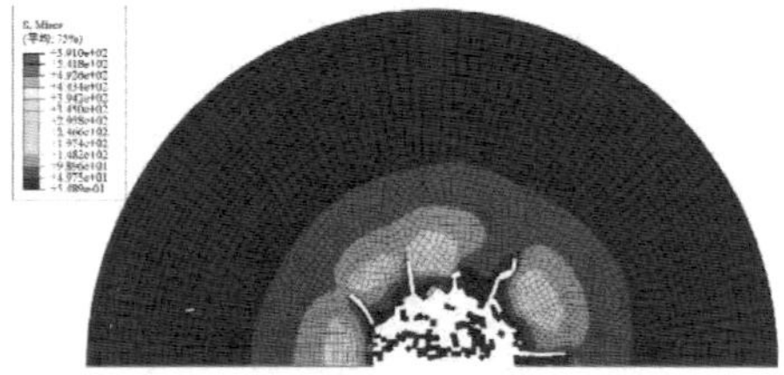

(c) v=20 m/s

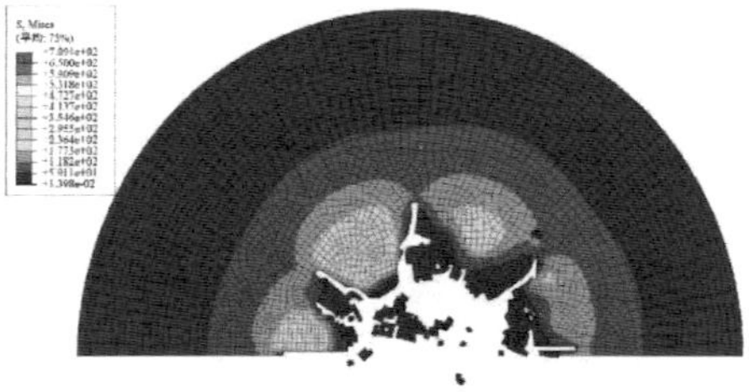

(d) v=30 m/s

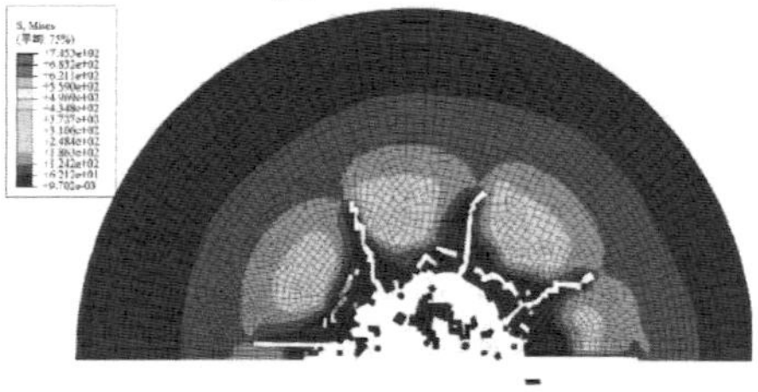

(e) v=40 m/s

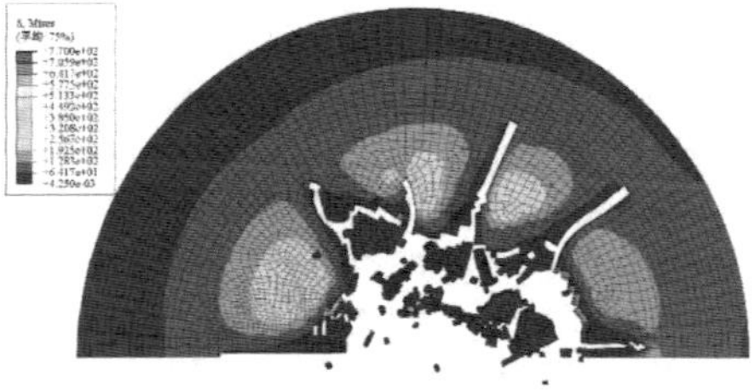

(f) v=50 m/s

Figure 5. The final destruction effect of the scale layer after impact of different blasting velocity

There are two types of failure unit. One is the unit direct failure, which is that the strain of the scaly cell itself is critical to the critical strain (in this model, the partially deactivated unit has been removed from the model after the calculation).

The second type is the failure of the scale unit group, these unit groups due to the failure of the surrounding units and the formation of a "island", although these unit group stress/strain does not reach the critical strain value, but in the subsequent projectile impact or matrix deformation and other external forces Under the action of the "island" as a whole and the strip out of the matrix, as shown in Figure 5(f) in the blank area of the blue unit group.

5.2. Energy utilization rate

The define of energy utilization rate is ε:

$$\varepsilon = \frac{S}{E_k} \tag{3}$$

Here, S is the direct destruction area, EK is the energy of the projectile.

As can be seen from the formula, the energy utilization rate (ε) characterizes the area of the scale that one unit energy can be removed from.

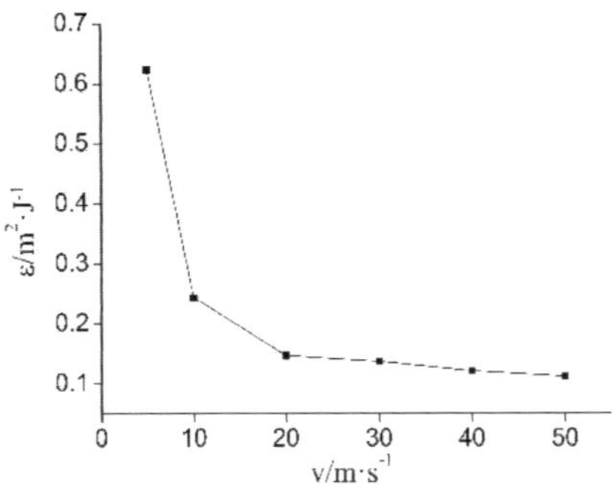

Figure 6. Energy utilization rate at different projectile impact speed

Figure 6 shows the relationship between energy utilization rate and projectile velocity. As can be seen from Figure 6, the energy utilization rate (ε) decreases as the projectile velocity increases.

On the actual descaling production line, the smaller the speed of the project, the slower the processing speed, which means, the time required to process a roll of steel is longer. With respect to various conditions, the ideal range of projectile impacting velocity is 20 m/s-30 m/s.

5.3 Deformation caused by body deformation

Figure 7 shows the stress cloud at different moments of the impact of the projectile at 25 m/s.

As can be seen from the Figure 7, as time goes, under the impact of the substrate more and more obvious deformation of the elastic-plastic is achieved, the stress quickly spreads among the surrounding area of radiation. Strip surface of the initial stress line is roughly circular, late was "star" to expand around.

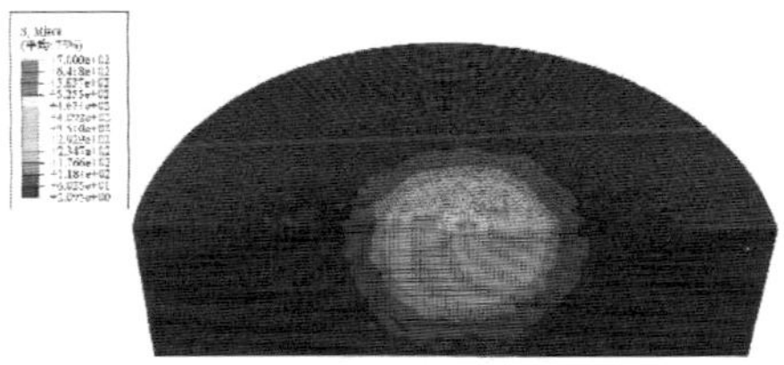

(a) $t=6\times10^{-7}$ s

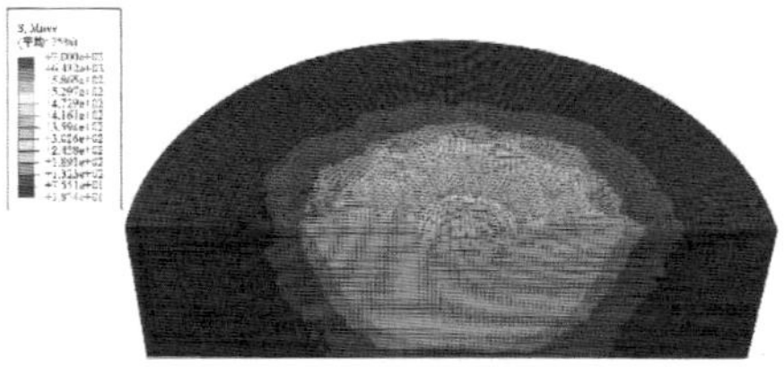

(b) t=8×10^{-7} s

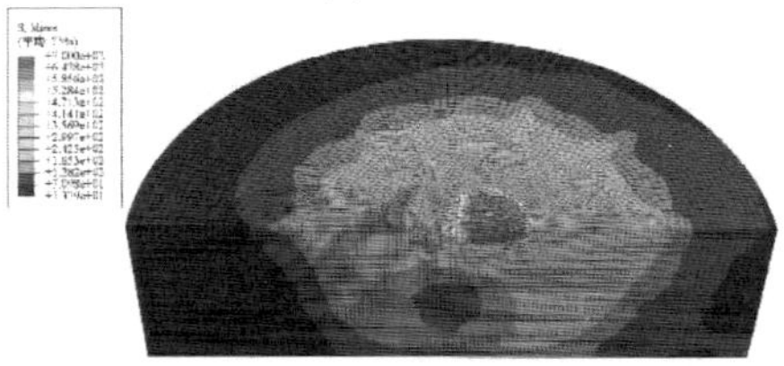

(c) t=1.2×10^{-6} s

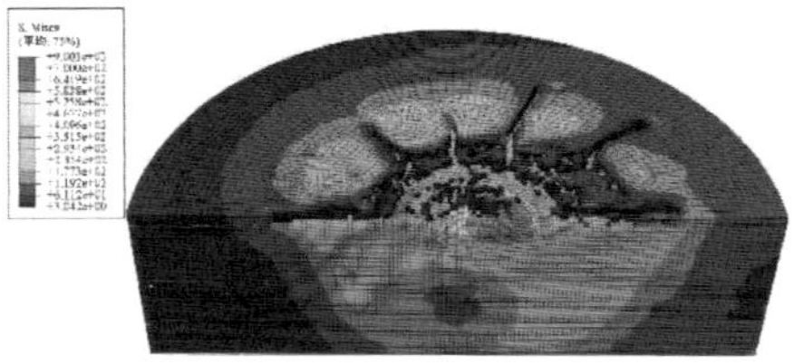

(d) t=1.2×10^{-6} s

Figure 7. Body stress at different time

Figure 7(c) and Figure 7(d) are body stress map and surface layer failure map at the same time (t = 1.2×10^{-6} s). After comparison of the two figures, the followings can be concluded:

(1).The better elastomeric body is rapidly exposed to the surroundings after impact, and the same size strain is transmitted to the surface layer (a tear effect) with the contact surface, which results in a scaly strain that is far from the direct impact zone, and fail to reach the damage value.

(2).As shown in Figure 7(c), since the iso-parametric lines of the late-stage steel strip are "star-shaped" outward, resulting in farther damage area spreading out as " whisker cracking" (figure d), "star-shaped" sharp corners of the stress line correspond to the "extension" of the failure zone of the scale.

If the "epitaxial zone" in Figure 7(d) is an indirect descaling zone, it is obvious that the scales in the area are easily destroyed by the subsequent impact of other pellets.

6. Conclusion

(1).Hot rolled strip Q235-A surface oxide scale is compact and has a good structure with the substrate, the thickness is about 10 μm, the critical damage strain is 0.08%.

(2).With the increase of the impact speed of the projectile, the damage effect of the scale layer is stronger. The fractured area of the scale has two parts, one is the direct impact damage area, and he other is the indirect tearing damage area caused by the body deformation. The indirect tearing is due to the large deformation of the iron body due to the impact of the iron body, and the strain is transferred to the surface oxide scale by means of the contact surface, causing the scale of the farther region.

(3).The bigger the shot speed, the lower the energy utilization rate. Taking into account the actual production situation, a reasonable impact speed should be 20m/s-30m/s.

References

[1] Gleeson B, Hadavi S M M and Young D J 2000 *Isothermal transformation behavior of*

IOP Conf. Series: Materials Science and Engineering **282** (2017) 012011 doi:10.1088/1757-899X/282/1/012011

thermally-grown wüstite[J]. Materials at High Temperatures **17(2)** pp 311

[2] Schwerdtfeger K and Shunxin Z A 2003 *contribution to scale growth during hot rolling of steel[J]. Steel research international* **74(9)** pp 538

[3] Dong Hanjun and Wang Yinjun 2009 *Factors Influencing Pickling Quality and Speed of Oxide Scale on Hot Rolled Strip[J]. Materials for Mechanical Engineering* **33(3)** pp 83 (in Chinese)

[4] Khoo B C, D. X U and Yuen W Y D, 2014 *et al. A Computer Analysis of Turbulent Flow of Acid in the Picking of Steel Strips[C]// Recent Advances in Computational Science and Engineering - the International Conference on Scientific and Engineering Computation* pp 171-75.

[5] Melster S and Pargmann A 1995 *Stretch bending levelers in pickling lines[J]. Iron Steel Eng.* **72(2)** pp 37-41

[6] Voges K C and Mueth A R 2012 *Method of producing rust inhibitive sheet metal through scale removal with a slurry blasting descaling cell: US* US8128460[P]

[7] Chaudhuri S K and Rolls R 1977 *Fracture mechanisms in oxide scale on iron during substrate deformation[J]. Journal of Materials Science* **12(11)** pp 2303

[8] Evans H E 1989 *Cracking and spalling of protective oxide layers[J]. Materials Science and Engineering: A* **120** pp 139

[9] Robertson J and Manning M I 1990 *Limits to adherence of oxide scales[J]. Materials Science and Technology* **6(1)** pp 81

[10] Nagl M M and Evans W T 1993 *The mechanical failure of oxide scales under tensile or compressive load[J]. Journal of Materials Science* **28(23)** pp 6247

[11] Gao Z and Bu Y 2015 *Research on the wet shot blasting technology of strip steel surface topography and finite element simulation[C]// International Conference on Materials, Environmental and Biological Engineering*

[12] Ren Yujun, Ge Guojun and Li Jun 2014 *Mechanism Analysis about Breakdown and Feed of Slurry Abrasive in Wet Blasting. [J]. Metallurgical Equipment* **2** pp 013

[13] Chen R Y and Yeun W Y D 2003 *Review of the high-temperature oxidation of iron and carbon steels in air or oxygen[J]. Oxidation of metals* **59(5-6)** pp 433

[14] Hibbit K 1995 *Abaqus/Standard Users Manual[J]* i–iii

[15] Kanninen M F, Popelar C A and Saunders H 1985 *Advanced fracture mechanics[M]// Advanced fracture mechanics /. Oxford University Press* pp 419-20

[16] Johnson G R and Cook W H. A 1983 *constitutive model and data for metals subjected to large strains, high strain rates and high temperatures[C]//Proceedings of the 7th International Symposium on Ballistics* **21** pp 541-47.

[17] Lin Li, Zhi Xudong and Fan Feng, 2014 *et al. Determination of parameters of Johnson-Cook models of Q235B steel[J]. Journal of Vibration and shock* **33(09)** pp 153-58

[18] Torres M and Colás R A 2000 *model for heat conduction through the oxide layer of steel during hot rolling[J]. Journal of Materials Processing Technology* 1**05(3)** pp 258-63

Analysis on the cost structure of product recall for reverse supply chain

Feng Yanhua[1,2], Xia Xuhui[1,2,*] and Yang Zheng[1,2]

[1]Key Laboratory of Metallurgical Equipment and Control Technology of Ministry of Education, Wuhan University of Science and Technology, 430081 Wuhan, China;
[2]Hubei Key Laboratory of Mechanical Transmission and Manufacturing Engineering, Wuhan University of Science and Technology, 430081 Wuhan, China;

[*] Corresponding author : xiaxuhui@wust.edu.cn

Abstract. The research on the reverse supply chain of product recall mainly focused on the recall network structure, logistics mode and so on. In this paper, when product recall and supply channel are fixed, the specific structure and function expression of cost are analyzed according to the peak season and off-season of recall activities, and whether the assembly manufacturer, supplier and recyclers are cooperated situation, respectively, to build the total cost structure of the function model. Finally, the model is validated correctly through the automotive industry and the electromechanical industry.

1. Introduction

The existing product recall research mainly focuses on the conditions the cooperation mechanism can be achieved, the different pricing strategies for effective measurement and the distribution of benefits, etc. In the supply chain management, in order to achieve the rapid response to the needs of the middle and lower reaches of the node, and continue to reduce the overall inventory level of the supply chain, many scholars found the problem in the system is particularly prominent by assembly to order in the supply chain collaboration[1-2]. There are two major difficulties affecting the manufacturer, one is the customer's needs in constant change, the other is the uncertainty of the delivery of the manufacturer. In the process of reverse supply chain, manufacturers supply situation on the downstream manufacturing enterprises processing and storage will costs more prominent because of the changing for the time and quantity. Therefore, the overall benefits of remanufacturing optimization has an important role according to the reverse supply chain raw material supply characteristics. Xia Xuhui believed that the general concept of reverse supply chain includes the basic concepts of reverse supply chain, the path of reverse supply chain, the open-loop and closed-loop characteristics of reverse supply chain, the characteristics of reverse supply chain, the meaning of reverse supply chain and the supply chain of reverse supply chain strategic[3]. He put forward the reverse supply chain is the whole recycling process that the market waste products back to the original manufacturer or remanufacturing to disassemble, and then create a new product back to the market.

The reverse supply chain research focuses on the recovery of old commodities, including the reverse supply chain coordination of the main incentives, pricing mechanism, recycling products remanufacturing and so on[4]. In recent years, the recall of the incident is more frequent, but the research on product recall is relatively small. Reverse supply chain for product recall contains too many unstable factors. Firstly, the quantity of product recall can not be estimated; Secondly, product recall quality is uncertain, that is, recall products meet remanufacturing processing of the standard conditions can not be estimate; Next, product recall recovery time is unknown. So, these uncertainties present a serious challenge for the product recall for reverse supply chain system. In this paper, the replacement of parts and components for product recall activities as an example, the co-supply problem will be discussed between the assembly manufacturer, suppliers and recyclers, and the cost structure and its model for product recovery reverse supply chain will be further studied.

2. Model hypothesis

In the product recall activities, the cost of product recovery reverse supply chain can be reduced effectively by the management of replacement part of the delivery time and the rapid sharing of product information. Due to factors such as production capacity constraints and the bottleneck of the replacement part of the assembly manufacturer delivery time will have a greater impact. But the use of methods to shorten the delivery time of the supplier may increase the storage costs and give the enterprise an unnecessary burden. The timing of supply of replacement parts and the control of supply volume are the key factors that affect the delivery time of the assembly manufacturer and reduce the cost of recall activities. The uncertainty of the recall quantity is the core information of the whole recall activity. The reference model hypothesis is as follow[5]:

1) In the production of reverse supply chain, all the maintenance activities of the assembly manufacturer for the recalled product are the replacement of the replacement part A and the recovery part B. The supplier A is responsible for providing the replacement part A, the recycler B is responsible for recycling B, the two originals are assembled by the assembly manufacturer and returned to the customer. The quantity ratio of the replacement part A and the part B is 1: 1, and the replacement part A supply shortage will directly lead to fracture of production chain after the recall, thus affecting the entire recall activities;

2) In the production of reverse supply chain, the assembly manufacturer is ordered by the Q_B to the supplier A according to the quantity Q_B of the recalled product. The supply ratio of the supplier A and the recovery part B is α: 1, the supply quantity is Q_A, The Q_B of the recovery of the quantity B is subject to the normal distribution, that is, $Q_B \sim N(\mu, \sigma^2)$, and the assembly manufacturer using continuous inspection (Q,R) strategy;

3) In order to ensure the efficiency of the processing activities to win the trust of the customer, the recycler B will return the product to the assembly manufacturer in a very short period of time after the customer returns the product. The recycler B does not generate the storage cost, that is Recycling B storage cost $D_B=0$;

4) In order to save costs and the rapid operation of the reverse supply chain, the recycler B often choose near the processing enterprises to build sites, and even some recyclers will be located in the processing enterprises at the recovery point, so the cost of transport B can be ignored, that is Recycling B transport costs $T_B=0$;

5) Assembly manufacturer with sufficient processing capacity of the recall products for maintenance, parts replacement and other activities, will not produce the recall of the product accumulation;

6) Assembly manufacturer and recycler B in the reverse supply chain behavior are completely rational, that is, in accordance with the principle of the lowest cost and maximize the principle of efficiency to implement the decision.

3. Reverse supply chain supply cost composition

3.1. Symbol Description

Q_A: The supply of replacement part A ; Q_B: The supply of recovery part B ;
α: Ratio of replacement part A to recovery part B ; q: Unit shortage cost of replacement part A ;
C_A: Unit storage cost of replacement part A ; C_B: Unit storage cost of recovery part B ;

P_A: Unit purchase price of replacement part A ;
Pr: Unit compensation amount of recovery part B ;
R_A: Unit reorder cost of replacement part A ;
T_B: Transportation cost of recovery part B ;
D_B: Storage cost of recovery part B ;
S_A: Storage cost of replacement part A ;
S_B: Storage cost of recovery part B ;
L: Shortage cost ;
G: Purchase cost;
P: Recovery compensation cost;
Z: Reorder cost.

3.2. The supply cost of the assembly manufacturer at peak season

Such recall behavior has a relatively obvious recall season. In the early stages of product recall information release, there will be a recall boom, that is recall peak season, terminal market demand is volatile and it is often difficult to obtain relatively accurate predictions.

1) Storage cost of replacement part A

Since replacement part A need to be procured from other production sectors, a certain amount of replacement pieces A will be kept in advance in order to ensure a relatively stable production operation before the arrival of the recall season. A and B components of the assembly ratio of 1: 1, so there may be the assembly manufacturer reserves more replacement A before the recall peak season (i.e.α> 1). The storage cost of replacement part A is

$$S_A=C_A\cdot Q_B\cdot(\alpha-1) \tag{1}$$

2) Storage cost of recovery part B

Since the ratio of part A and B is 1: 1, and the number of products recalled in the reverse supply chain is large. Therefore, when the supply of the replacement part A is insufficient (i.e.α <1),it leads to the assembly manufacturer reserves a certain amount of recovery part B.The storage cost of recovery part B is

$$S_B=C_B\cdot Q_B\cdot(1-\alpha) \tag{2}$$

3) Shortage cost

Due to the error in the forecasting information of the assembly manufacturer and the uncertainty of the production capacity of the replacement part supplier, there may be a shortage of replacement parts A, resulting in an out-of-stock cost of the assembly is

$$L=q\cdot Q_B\cdot[1-\min(\alpha, 1)] \tag{3}$$

4) Purchase cost

By the actual supply of the replacement part A, it can be concluded that the purchase cost of the assembly manufacturer is

$$G=P_A\cdot\alpha\cdot Q_B \tag{4}$$

5) Recovery compensation cost

In order to promote the active participation of consumers in the recall activities, improve consumer satisfaction with business services, enterprises will be a certain recovery compensation in the provision of recall services.you can get the recovery compensation cost

$$P=Pr\cdot Q_B \tag{5}$$

6) Reorder cost

As the estimated recovery is less than the final recall of the number of products, the assembly manufacturer may be in the process of adding additional purchase A to meet the needs of consumers recall, for which the purchase cost

$$Z=R_A\cdot Q_B\cdot[1-\min(\alpha, 1)] \tag{6}$$

3.3. The supply cost of the assembly manufacturer at off- season

After the peak period of the recall, the number of product recalls in the reverse supply chain will gradually decrease and tend to be gentle. At the off-season, the demand for the market is close to that of the assembly manufacturer. Therefore, enterprises can reasonably predict the level of demand for recovery part B, and in order to ensure the smooth progress of the recall, the assembly manufacturer may reserve a certain number of replacement part A.

1) Storage cost of replacement part A

A and B components of the assembly ratio of 1: 1, so there may be assembly (manufacturing) in the recall of the off-season reserve more replacement A (α> 1), the reserve cost is

$$S_A=C_A\cdot Q_B\cdot(\alpha-1) \quad (7)$$

2) Recovery compensation cost

In order to promote the customer to take the initiative to recall activities to improve consumer satisfaction with the enterprise services, enterprises in the provision of recall services, often also a certain recovery compensation, the recovery of the recovery costs can be get, that is

$$P=Pr\cdot Q_B \quad (8)$$

3) Purchase cost

By the actual supply of the replacement part A, it can be concluded that the purchase cost of the assembly manufacturer is

$$G=P_A\cdot \alpha\cdot Q_B \quad (9)$$

4. Total cost of reverse supply chain

4.1. Models that are non-coordination between enterprises

In this reverse supply chain model, the assembly manufacturer only share the expected information on the recalled product with the recycler of the recovery part B, and does not pay attention to the sharing of information with the replacement parts supplier. At the same time, the total cost of the assembly manufacturer during the peak season is

$$W_1=S_A+S_B+L+G+P+Z=$$
$$C_A\cdot Q_B\cdot(\alpha-1)+C_B\cdot Q_B\cdot(1-\alpha)+q\cdot Q_B\cdot[1-\min(\alpha,1)]+P_A\cdot \alpha\cdot Q_B+P_r\cdot Q_B+R_A\cdot Q_B\cdot[1-\min(\alpha,1)] \quad (10)$$

The total cost of the assembly manufacturer during the peak season is

$$W_2=S_A+P+G=C_A\cdot Q_B\cdot(\alpha-1)+Pr\cdot Q_B+P_A\cdot \alpha\cdot Q_B \quad (11)$$

Therefore, at the peak season the expected value of total cost is

$$E(W_1)=E(S_A)+E(S_B)+E(L)+E(G)+E(P)+E(Z)=$$
$$C_A\cdot Q_B\cdot\int_{\alpha\geq1}(\alpha-1)f(\alpha)d\alpha+C_B\cdot Q_B\cdot\int_{\alpha\leq1}(1-\alpha)f(\alpha)d\alpha+q\cdot Q_B\cdot\int_{\alpha\leq1}(1-\alpha)f(\alpha)\,d\alpha$$
$$+P_A\cdot Q_B\cdot E(\alpha)+P_r\cdot Q_B+R_A\cdot Q_B\cdot\int_{\alpha\leq1}(1-\alpha)f(\alpha)\,d\alpha \quad (12)$$

At the off- season the expected value of total cost is

$$E(W_2)=E(S_A)+E(P)+E(G)=C_A\cdot Q_B\cdot\int_{\alpha\geq1}(\alpha-1)f(\alpha)d\,\alpha+P_A\cdot Q_B\cdot E(\alpha)+P_r\cdot Q_B \quad (13)$$

4.2. Models that are coordination between enterprises

When the supplier, the recycler and the assembly manufacturer share information and effective collaboration, the parties can be clear each other's supply quantity and delivery time.They will reduce the holding cost of replacement part or recovery part and reduce the shortage cost and reorder cost due to a shortage of replacement part,thereby reducing the cost of the entire reverse supply chain. At this point, during the peak season of the recall, the total cost of the assembly manufacturer is

$$W_1'=L+G+P=q\cdot Q_B\cdot[1-\min(\alpha,1)]+P_A\cdot \alpha\cdot Q_B+P_r\cdot Q_B \quad (14)$$

During the off season, the total cost of the assembly manufacturer becomes

$$W_2'=P+G=P_r\cdot Q_B+P_A\cdot \alpha\cdot Q_B \quad (15)$$

Therefore, the peak season at the same time expect the total cost of manufacturing is

$$E(W_1')=E(L)+E(G)+E(P)=q\cdot Q_B\cdot\int_{\alpha\leq1}(1-\alpha)f(\alpha)d\,\alpha+P_A\cdot Q_B\cdot E(\alpha)+P_r\cdot Q_B \quad (16)$$

The expected total cost of the off-season association is

$$E(W_2')=E(P)+E(G)=P_A\cdot Q_B\cdot E(\alpha)+P_r\cdot Q_B \quad (17)$$

Compared with the total cost of manufacturing in different times, it can be seen that the establishment of information channels, information sharing and co-production can effectively reduce the production cost of reverse supply chain, to avoid product accumulation or lack of supply situation.

5. Example verification

The model is validated through the automotive industry and the electromechanical industry. Assuming that in a certain area, a car repair shop is responsible for servicing the recall car in the area, and the recall is in line with the set model. It is known that the density function of random variable α is f (α). It is assumed that α is uniformly distributed U (a, b) | (a=0.8,b=1.2) at the peak season and uniformly distributed U (a, b) | (a=1.1, b=1.5) at the off-season; In another area, a company flagship store appliance repair service department is responsible for the recall refrigerator door activities. It is assumed that α is approximate uniformly distributed U (a, b) | (a=0.7,b=1.1) at the peak season and approximate uniformly distributed U (a, b) | (a=1, b=3) at the off-season. The cost of each link remains the same during the recall season, and the relevant calculation parameters are shown in Table 1, and the the assembly manufacturer is processed at a ratio of 1: 1 between Q_A and Q_B. The relevant calculation results of the peak season and the off-season are obtained by generating the random variable of 10000 pairs of α, which is the random distribution of α , and the supply level of the replacement part supplier and the recycler. The relevant calculation results are shown in Table 2.

Table 1. The relevant calculation parameters

Business Model	q	C_A	C_B	P_A	P_r	R_A
Car	50	10	15	20	10	30
Refrigerator	80	20	25	40	20	50

Table 2. The relevant calculation results

Business Model	Recall at peak season				Recall at off-season			
	f(a)	E(a)	$E(W_1)$	$E(W_1')$	f(a)	E(a)	$E(W_1)$	$E(W_1')$
Car	2.5	1	$35.25Q_B$	$32.5Q_B$	2.5	1.3	$39Q_B$	$36Q_B$
Refrigerator	2	0.9	$71.95Q_B$	$63.2Q_B$	0.5	2	$120Q_B$	$100Q_B$

Figure 1 shows the impact of the different callbacks on the total cost expectancy in the automotive industry, reflecting the total cost expectations and synergies between the recalled peak season and the recalled off-season replacement part supplier, recycler, and assembly manufacturer changes in cost expectations; Figure 2 shows the impact of the different recovery volumes on the total cost expectancy in the refrigerator industry, reflecting the total cost of the recall and the off-season replacement parts suppliers, flagship stores and appliance repair services, and the total cost of synergies The trend of change of expectation. It can be seen that the total cost of the firm in the cooperative state is lower than in the non-cooperative state, so the above model is assumed to be true. When the parts suppliers (replacement parts suppliers), car repair shop (flagship store) and 4S shop (home appliance repair service department) share information and effective collaboration, the parties can be clear each other's supply quantity and delivery time.They will reduce the holding cost of replacement parts or recovery part and reduce the shortage cost and reorder cost due to a shortage of replacement part.

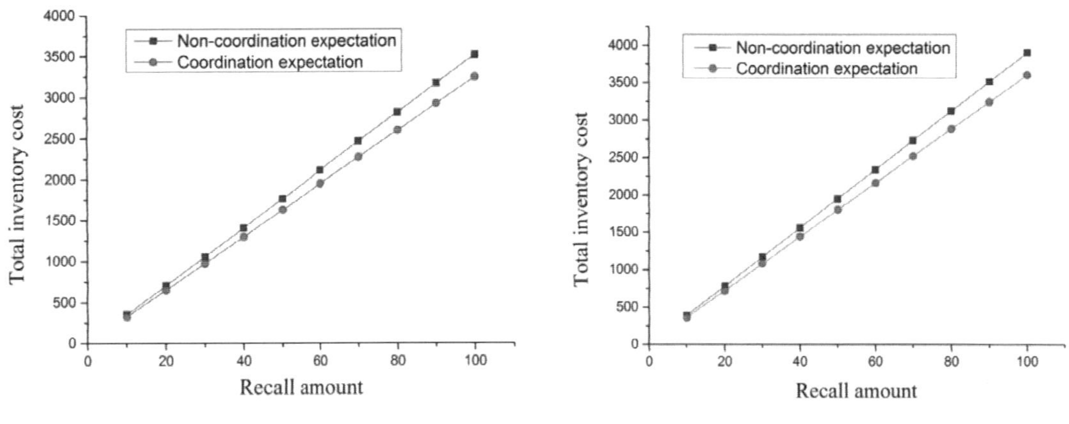

(a) Recall at peak season (b) Recall at off-season

Figure 1. The impact of different callbacks on total cost expectations in automobile industry

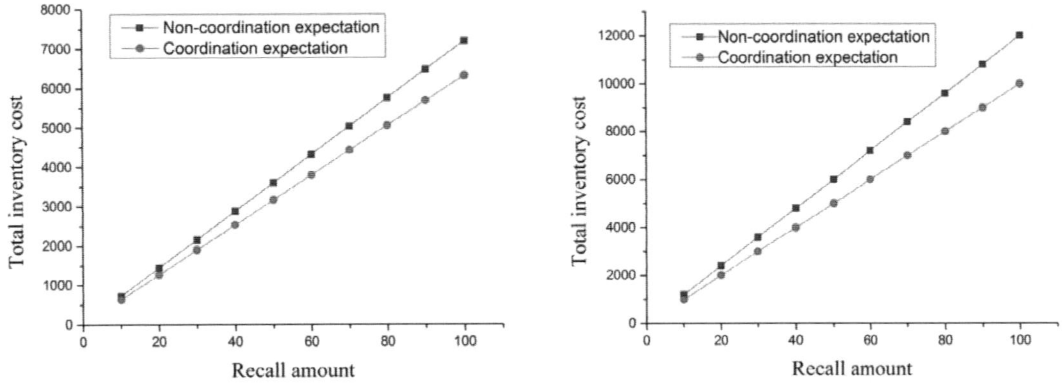

(a) Recall at peak season (b) Recall at off-season

Figure 2. The impact of different callbacks on total cost expectations in refrigerator industry

6. Conclusion

In this paper, in the case of product recall channel and supply channel fixed, the specific structure and function expression of cost are analyzed according to the peak season and off-season of recall activities, and whether the assembly manufacturer, supplier and recycler are cooperated situation, respectively, to build the total cost of the function model.

1) When the assembly manufacturer, suppliers and recyclers share information and co-supply, the three can clear each other's supply and delivery time, it will effectively reduce production costs.

2) The recall activities have a significant difference at the recall peak season and off- season. At the peak season, the demand in the terminal market fluctuates greatly, and it is difficult to obtain relatively accurate forecast results. Compared with the recall of the off-season, due to the recovery of the reserves, replacement part out of stock, re-purchase of replacement part to bring new costs.

3) Assume that consumer product returns are subject to uniform distribution, which is somewhat different from the reality. The article does not consider whether the replacement part supplier will be able to meet the re-order of the assembly manufacturer in the peak season. In the follow-up study, gradually reduce the restrictions, more thinking the replacement part supplier and recycler in advance delivery, etc. So that the relevant enterprises to better participate in the collaborative process, play a greater role in collaborative supply.

Acknowledgments

This work was supported by the National Natural Science Foundation of China (No. 714711434).

References

[1] C Prahinski and C Kocabasoglu 2006 *Omega* **34** pp 6
[2] G LI, S Ma and Z H Wang 2011 *et al. Journal of Mechanical Engineering* **47** pp 20
[3] X H Xia, F Liu 2005 *Journal of Mechanical Engineering* **41** pp 4
[4] H Wang 2006 *Logistics Science and Technology* **29** pp132
[5] Y F Yu, C C Yu and J J Pan 2016 *Computer Simulation* **33** pp 8

IOP Conf. Series: Materials Science and Engineering **282** (2017) 012013 doi:10.1088/1757-899X/282/1/012013

Size effects in micro rolling of metals

Jingwei Zhao[a], Haibo Xie, Haina Lu and Zhengyi Jiang[b]

School of Mechanical, Materials and Mechatronic Engineering, University of Wollongong, NSW 2522, Australia

[a] E-mail: jwzhaocn@gmail.com [b] E-mail: jiang@uow.edu.au

Abstract. Size effects are a unique characteristic with miniaturisation, and may cause a number of unexpected problems in micro rolling of metals in the aspects of mechanical behaviour, tribology, and scatter of material behaviour. This paper has introduced three kinds of micro rolling technologies, including micro ultrathin strip rolling, micro flexible rolling and micro cross wedge rolling, and analysed the size effects appeared in each of the micro rolling process. This paper will be helpful for understanding the problem of size effects, and can contribute to the manufacturing of high quality micro metallic products by minimising the impact of size effects in micro rolling with optimised processing parameters.

1. Introduction

Product miniaturisation, which aims to reduce product weight, volume, cost and pollution, is a global trend for facilitating product usage and enabling product functions to be implemented in microscale geometries. Increased demands for miniaturised products have greatly accelerated the development of microforming technology, one of the most popular micromanufacturing processes for the fabrication of very small metallic parts in the submillimeter range [1]. The fabricated micro products have tremendous applications in diverse areas including micro-electromechanical systems (MEMS), micro fluidics systems (MFS), medical devices, precision equipment, and communication devices. Microforming is applicable to a wide range of materials, and can be used to economically fabricate metallic micro products with complex geometries, excellent dimensional tolerances, required mechanical properties and improved surface quality [2].

In metalworking, rolling is metal forming process in which workpiece is passed through one or more pairs of rolls or featured tools to reduce the thickness or to form specifically shaped products. Micro rolling, as the name suggests, is one of the microforming methods. In micro rolling, the dimensions of workpiece along the deformation direction are in the submillimeter range, and the rolling deformation is processed in microscale. Traditionally, micro rolling refers to micro ultrathin strip rolling by which small and ultrathin strip is flat-rolled to reduce the thickness and to make the thickness uniform. In micro ultrathin strip rolling, the width and thickness of the strip being rolled are normally in the domains of lower millimetre and submillimetre, respectively. With the development of rolling and microforming technology, other micro rolling methods, including micro flexible rolling [3, 4] and micro cross wedge rolling [5], have also been successfully developed.

Micro rolling can be employed to manufacture net-shape or near net-shape micro products with high quality. One direct way to realise micro rolling process is by scaling down the rolling system from the macroscopic to the mesoscopic domain. Unfortunately, rolling theories and technologies established in the macro world cannot be simply scaled down to the micro world, because it is impossible to scale down all parameters in the micro rolling process according to the theory of similarity due to the size effects which are well known as a unique characteristic with miniaturisation. Size effects occur due to the fact that the ratio among all decisive features cannot be kept a constant according to the process requirements, and may cause a number of unexpected problems in micro rolling process in the aspects of mechanical behaviour, tribology, and scatter of material behaviour. Therefore, the size effects in micro rolling have to be identified in order to manufacture high quality micro products that meet the high demands.

This paper aims to characterise the size effects in micro rolling of metals. In the following sections, three micro rolling technologies, including micro ultrathin strip rolling, micro flexible rolling and micro cross wedge rolling (MCWR), will be respectively introduced followed by an analysis on the size effects appeared in each of the micro rolling process.

2. Size effects in micro ultrathin strip rolling

Ultrathin strip manufactured by micro rolling has a wide application in electronic and instrument industries, and its production has always been of interest to the manufacturers and researchers in the area of metal rolling. The principle of micro ultrathin strip rolling is the same as that of conventional flat rolling, and the rolling process can be schematically illustrated in Figure 1. Differently, the deformation of ultrathin strip in micro rolling is affected by size effects.

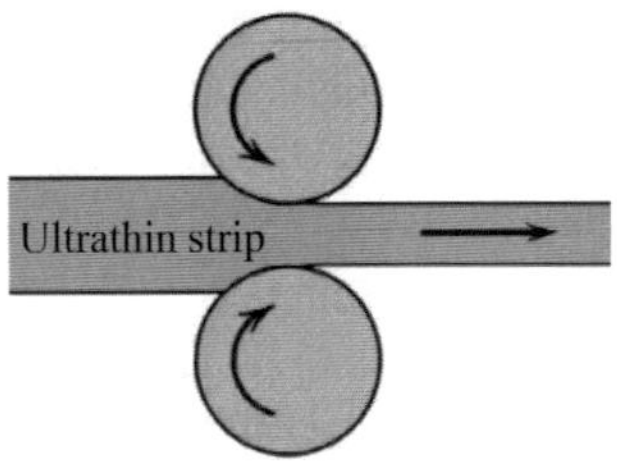

Figure 1. Schematic illustration of micro ultrathin strip rolling.

With miniaturisation of the rolled ultrathin strip, the ratio of grain size to the characteristic dimension will be changed. When the strip's thickness is kept constant but the grain size is increased, the yield strength of the strip will be decreased according to the Hall-Petch effect [6, 7]. Similarly, the yield strength will also be reduced when the strip's grain size is kept unchanged but the thickness is reduced based on the surface layer model [2]. The above both cases may lead to reduced flow stress in rolling. With increasing the grain size or reducing the strip's thickness, the number of grains through the thickness becomes so small that the strip's behaviour will be characterised by only a few grains. As the grains possess different sizes, shapes and orientations, each grain will play a significant role in plastic deformation in micro ultrathin strip rolling.

The coefficient of friction (COF) will increase with miniaturisation in micro ultrathin strip rolling. An increase in COF induces increased rolling force, intermediate force and edge contact force. The work roll edge contact can increase work roll edge wear, which will shorten the work roll service life. The strip profile can be improved when the COF increases [8]. The edge kiss of work rolls can also improve the strip shape when no work roll bending force is applied. The edge kiss force between the two work rolls will increase when the COF increases, which can improve the shape of ultrathin strip in rolling [9]. When suitable friction conditions along the longitudinal and transverse directions are applied to the micro rolling process, a good shape and profile of ultrathin strip can be obtained [10]. Surface roughness affects the friction behaviour, stress distribution, and surface quality, which will become more significant with the scaling down of strip's thickness. The surface roughness will decrease with an increase in pass reduction due to the reprinting or interaction between the rolls and the rolled ultrathin strip.

In micro ultrathin strip rolling, lubrication plays an important role in the cooling of rolled material, reducing of rolling force, decreasing of wear of both rolls and rolled material, and lowering of friction. When oil-based lubricant is applied to micro ultrathin strip rolling, the oil film thickness decreases with the increase of pass reduction and the decrease of rolling speed. When the oil film thickness is increased, rolling force will be reduced, the surface of the product, however, will become dull. Lubricating condition is remarkably influenced by roll-coating formation, especially by using smooth surface of rolls. For micro ultrathin strip rolling, the most preferable lubricant is emulsion [11]. The size effects on lubricant friction behaviour can be explained by the model of open and closed lubricant pockets, also known as dynamic and static lubricant pockets, respectively. Downscaling affects the ratio of open to closed lubricant pockets. As the topography is not scaled down with geometrical similarity, there will be a region of constant width where open lubricant pockets become effects. The open lubricant pockets have a connection to the edge of the surface and cannot keep the lubricant. When scaling down the strip's size, the contact area in rolling is reduced while the size of single topography features remains approximately constant, leading to increased ratio of open lubricant pockets. As a result, the friction and rolling force will increase in micro ultrathin strip rolling as compared to the conventional rolling process.

3. Size effects in micro flexible rolling

Micro flexible rolling is a novel microforming method for the production of ultrathin strip with longitudinal thickness variation. In micro flexible rolling, a gap control system is applied to adjust the roll gap online deliberately during rolling operations, and to manufacture the thickness transition along the rolling direction, as schematically illustrated in Figure 2. For achieving periodical changes in the roll gap, the top and bottom work rolls can move simultaneously in opposite directions, or only the top work roll moves, along the normal direction of the rolled strip. Adjustment of roll gap by vertical movement of the top work roll is commonly found in micro flexible rolling of metals [3].

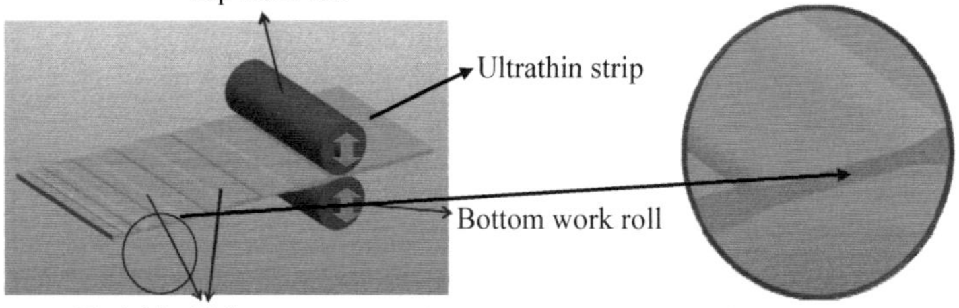

Figure 2. Schematic illustration of micro flexible rolling.

Micro flexible rolling differs from conventional flexible rolling because size effects that characterise the micro world are involved in the micro flexible rolling process. Significant springback problem occurs during micro flexible rolling process due to size effects, which has a great impact on the rolled products. In conventional flexible rolling, springback is negligible due to the absence of size effects, and the quality of the rolled sheets can be well controlled [12]. In micro flexible rolling, however, even a small springback of the rolled strip may cause big problem of dimensional accuracy due to the small deformation of ultrathin strip and the high requirement of dimensional tolerance. It is found that the average springback shows a decreasing trend with the increase of front and back tensions in case of micro flexible rolling of ultrathin aluminium alloy strips. For strips with thickness values of 100, 250 and 500 μm, more front and back tensions play a more crucial role in eliminating the springback despite the change of reduction [4].

The ratio of strip's thickness (T) to grain size (D) has a great effect on springback during micro flexible rolling. The ratio (T/D) decreases with scaling down the strip's thickness or increasing the grain size of ultrathin strip that has a constant thickness. In micro flexible rolling, the decrease of strip's thickness or the increase of grain size will result in a big springback, and the scatter of springback will become more serious with the decrease of T/D. This is because the grain number in the thickness direction drops due to the decrease of T/D, and the springback depends mainly on the mechanical behaviour of these limited grains. When the value of T/D increases, the effect of a single grain on the springback becomes small. As the springback of strip and the scatter of springback are reduced with the increase of T/D, high accuracy ultrathin strip with varying thickness can be successfully manufactured by micro flexible rolling method [2]. It should be noted that the degree of springback for different kinds of materials, such as steels, aluminium alloys and copper alloys, is different due to the difference in mechanical properties of the materials. Therefore, the springback that is caused by size effects should be identified specifically in the micro flexible rolling of different materials.

Ultrathin strip with varying thickness and perfect shape, flatness and optimal thickness transition length has wide applications in manufacturing industry. As a novel microforming method, micro flexible rolling is still under development. As the thickness of workpiece in micro flexible rolling is in the submillimeter range, reduction will become difficult due to the existence of a limit on the thickness of workpiece which can be reduced on a given mill. This minimum gage phenomenon is the result of a complex interaction of effects between the rolls and the strip rather than a function of the strip behaviour or of the mill behaviour along [13]. As a result, control of the reduction of the transition length of the varying strip thickness zone, and as well as the strip shape, profile, and flatness for thin strip, will be challenges for researchers and manufacturers in micro flexible rolling. This requires development of advanced micro flexible rolling equipment with high stiffness rolls and high-precision control system that have the capacity to minimise the impact of size effects and realise rapid dynamic response to change the roll gap and roll speed when the rolling mill is running.

4. Size effects in micro cross wedge rolling

Cross wedge rolling (CWR) is a metal forming technology in which a cylindrical rod is plastically deformed into an axisymmetrical part by the action of wedge-shaped dies moving tangentially relative to the workpiece. Conventionally, CWR is an efficient metal forming technology applied mainly for the manufacturing of stepped axles and shafts, and has been proven to be an economical and reproducible means of producing preforms for forging operations. CWR has three key features, including higher productivity, higher material utilisation and better working environment, as compared to other forming methods. MCWR technology derives from the concept of CWR, with CWR being scaled down into the micro domain. There are generally five different variants of CWRs in which two or three wedges mounted on rolls or flat or concave plates of rolling mills are used [14]. In MCWR, flat wedges are widely used due to its fewer requirements for sophisticated tooling structures so that tool manufacturing is much easier compared to other kinds of wedges, such as two-roll wedges, three-roll wedges and two concave wedges etc. Figure 3(a) shows an example of stepped shaft micro product manufactured by MCWR technology with a pair of flat wedges that are made of tool steel (Figure 3(b)).

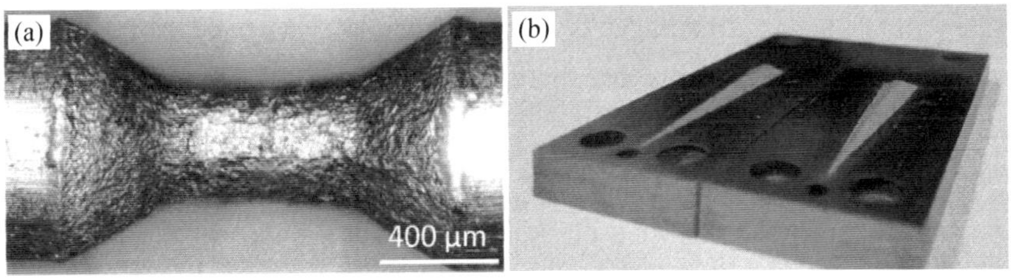

Figure 3. (a) Stepped shaft micro product manufactured by MCWR with (b) a pair of flat wedges.

Figure 4 schematically illustrates the MCWR forming process by using flat wedges, in which cross sections in five rolling stages have been shown: (I) knifing zone, (II) guiding zone, (III) stretching zone, (IV) sizing zone, and (V) unloading zone [15]. Wedges first cut into workpiece in its middle part and then, the V-shaped groove is generated along the entire circumference after the guiding zone. Subsequently, in the stretching zone the reduction of cross section is developed to the required width. Finally the geometrical dimensions of workpiece are refined during the sizing zone. The details of MCWR, including equipment, tools and process can be found elsewhere [2].

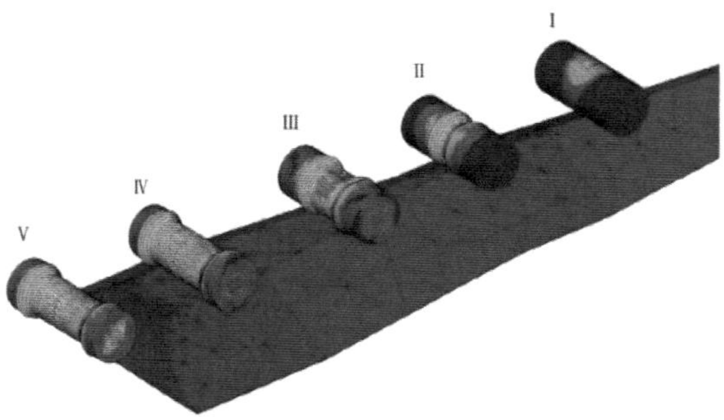

Figure 4. Schematic illustration of MCWR process consisting of five stages.

The original diameter of cylindrical rod being rolled in MCWR is generally in the submillimetre range. As a common problem in microforming, size effects also take place in the MCWR process of metals. Size effects have significant impact on the processing factors in MCWR, for instance, springback and surface topology of workpiece. In MCWR, the springback behaviour is related to the material's microstructure and deformation amount. The springback along both the axial and redial directions of manufactured product by MCWR as well as the springback scatter exhibit an increasing trend with the increase of grain size. Grain size has great effect on the strain and stress distribution of the workpiece processed by MCWR. Figure 5 shows the cross sections located in the middle of the cylindrical rods when rolling deformation of the rods is in the guiding stage (stage II illustrated in Figure 4). It can be seen that the continuous laminar distribution of stress and strain in the workpiece has been more significantly disturbed due to inhomogeneous mechanical properties with an increase of grain size.

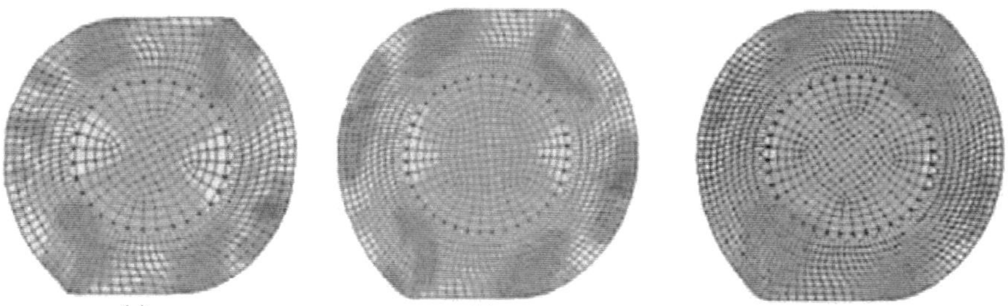

Figure 5. von Mises stress distribution on middle cross section of MCWR-processed workpieces with different grain sizes: (a) 248 μm, (b) 40 μm, and (c) 6 μm [15].

The springback of workpiece will become more significantly when the area reduction increases in MCWR. This is because that with the increase of rolling deformation, strain penetrates more deeply into the grains, resulting in activation of more dislocations and intertwinement of repelling dislocations which will release more inelastic springback in the unloading stage. Also, with miniaturisation, the relative ratio of the surface asperities of tools and blanks to the outside dimensions of workpiece becomes larger than that in the case of conventional macroscale process. As a result, the surface asperities may affect the frictional behaviour during MCWR process, and the processing characteristics and accuracy of the manufactured micro products [2].

In the presence of size effects, the distribution and evolution of strain and stress will become irregular in the processing of MCWR due to grain heterogeneity, which will induce inhomogeneous deformation. A homogeneous phenomenon of deformability of workpiece could be achieved at elevated temperatures, which could mitigate size effects and in turn lead to improvements in the edge quality of the products manufactured by MCWR due to the activation of more slip systems with thermal activation [5]. In order to take full advantage of MCWR technology, comprehensive

IOP Conf. Series: Materials Science and Engineering **282** (2017) 012013 doi:10.1088/1757-899X/282/1/012013

investigations on the influence of miniaturisation, especially size effects, on process, accuracy control, and product quality are still needed.

5. Conclusions

Micro rolling is one of the microforming methods, and has been under prompt development with the global trend towards miniaturisation in diverse areas. In this paper, three micro rolling technologies, including micro ultrathin strip rolling, micro flexible rolling and micro cross wedge rolling, have been introduced, and the size effects appeared in each of the micro rolling process have been analysed. Size effects characterise the micro world, and have significant impact on material behaviour in micro rolling and the quality of rolled products. Through understanding the mechanisms of size effects and proposing appropriate control strategies, high-quality micro metallic products can be successfully manufactured by using micro rolling technologies with optimised processing parameters.

References

[1] Z Jiang, J Zhao, H Lu, D Wei, K I Manabe, X Zhao, X Zhang and D Wu 2016 *Int. J. Mater. Form. DOI: http://dx.doi.org/10.1007/s12289-016-1317-4*

[2] Z Jiang, J Zhao and H Xie, 2017 *Microforming Technology: Theory, Simulation and Practice* (Elsevier)

[3] F Qu, Z Jiang, D Wei, Q Chen and H Lu 2017 *Int. J. Mech. Sci.* **123** pp 324

[4] F Qu, Z Jiang, H Lu 2016 *Int. J. Mech. Sci.* **105** pp 182

[5] H Lu 2013 *A Study on the Micro Cross Wedge Rolling of Metals* (University of Wollongong)

[6] E O Hall 1951 *Proc. Phys. Soc. B* **64** pp747

[7] N J Petch 1953 *J. Iron Steel Inst.* **174** pp 25

[8] Z Y Jiang and A K Tieu 2007 Wear **263** pp 1447

[9] Z Y Jiang, D Wei and A K Tieu 2009 *J. Mater. Process. Technol.* **209** pp 4584

[10] Z Y Jiang, S W Xiong, A K Tieu and Q J Wang 2008 *J. Mater. Process. Technol.* **201** pp 85

[11] H Xie, K I Manabe, T Furushima, K Tada and Z Jiang 2016 *Int. J. Adv. Manuf. Technol.* **82** pp 65

[12] O Engler, C Schafer, H –J Brinkman, J Brecht, P Beiter and K Nijhof 2016 *J. Mater. Process. Technol.* **229** pp 139

[13] H A Kuhn and A S Weinstein 1971 *J. Lubricat. Technol.* **93** pp 331

[14] Z Pater, A Gontarz and W Weroński 2006 *J. Mater. Process. Technol.* **177** PP 550

[15] Z Jiang, H Lu, D Wei, K Z Linghu, X Zhao, X Zhang and D Wu 2014 Procedia Eng. **81** PP 2463

Modeling a hierarchical structure of factors influencing exploitation policy for water distribution systems using ISM approach

Małgorzata Jasiulewicz-Kaczmarek[1,*], Ryszard Wyczółkowski[2] and Violetta Gładysiak[3]

[1]Poznan University of Technology, Faculty of Engineering Management, Strzelecka St. 11, Poland
[2]Silesian University of Technology, Faculty of Organization and Management, Roosevelta St. 26, Zabrze, Poland
[3]ZW-K Śmigiel Sp. Z o.o., dr Skarzyńskiego St. 6A, Śmigiel, Poland

[*] Corresponding author : malgorzata.jasiulewicz-kaczmarek@put.poznan.pl

Abstract. Water distribution systems are one of the basic elements of contemporary technical infrastructure of urban and rural areas. It is a complex engineering system composed of transmission networks and auxiliary equipment (e.g. controllers, checkouts etc.), scattered territorially over a large area. From the water distribution system operation point of view, its basic features are: functional variability, resulting from the need to adjust the system to temporary fluctuations in demand for water and territorial dispersion. The main research questions are: What external factors should be taken into account when developing an effective water distribution policy? Does the size and nature of the water distribution system significantly affect the exploitation policy implemented? These questions have shaped the objectives of research and the method of research implementation.

1. Introduction

Water distribution systems are one of the basic elements of contemporary technical infrastructure of urban and rural areas. It is a complex engineering system composed of transmission networks and auxiliary equipment (e.g. controllers, checkouts etc.), scattered territorially over a large area. From the water distribution system operation point of view, its basic features are: functional variability, resulting from the need to adjust the system to temporary fluctuations in demand for water and territorial dispersion, which on the one hand allows for the realization of the basic goal of supplying water to consumers; while on the other is a serious impediment to maintenance work (inspections, repairs, etc.) as it implies the need to move people and equipment necessary to carry out the work and to ensure proper communication. In the exploitation policy of the water distribution network, especially in the case of temporary fluctuations in the demand for water, even advanced IT techniques such as artificial intelligence, including neural networks, should be used. In Poland, research is being carried out on forecasting of the water distribution network load over different periods of time [1,2]. Environmental aspects are also important during the development (design) and operation of the water distribution network. The literature [3,4] can be used here.

It is important to note that water distribution networks are usually large assets, of the value significantly higher than the income of its operator. There is therefore no financial or technical capacity to reproduce a significant part of the network in a short period of time. Hence, it is important to adopt exploitation policy so that, by systematic actions of a small extent (compared to the size and value of the facility being used), not only maintain the whole system and keep it exploitation, but continuously adjust it to growing requirements (increasing number of customers, increase in the length of the water distribution network, changes in legal requirements, etc.).

The main research questions are: What external factors should be taken into account when developing an effective water distribution policy? Does the size and nature of the water distribution system significantly affect the exploitation policy implemented? These questions have shaped the objectives of research and the method of research implementation. The objectives of the research are: (1) To identify and rank the external factors (variables) influencing the exploitation policy of the water distribution network; (2) To establish the relationship among these identified variables (3) To develop a structural model using Interpretive Structural Modeling (ISM); (4) To carry out MICMAC analysis. The research included the exploitation policy of the small water distribution network in less urbanized areas and compared the importance of the criteria considered for this network and the typical urban network described by [5, 6].

The article is composed of five chapters. Chapter 2 describes the ISM methodology, defines the objective of the method, its implementation, and literature review of the ISM method. The next chapter is the research part. This chapter includes characteristics the research object and analysis of the factors influencing the exploitation policy of the presented water distribution network. The simplified scheme is illustrated in Figure 1.

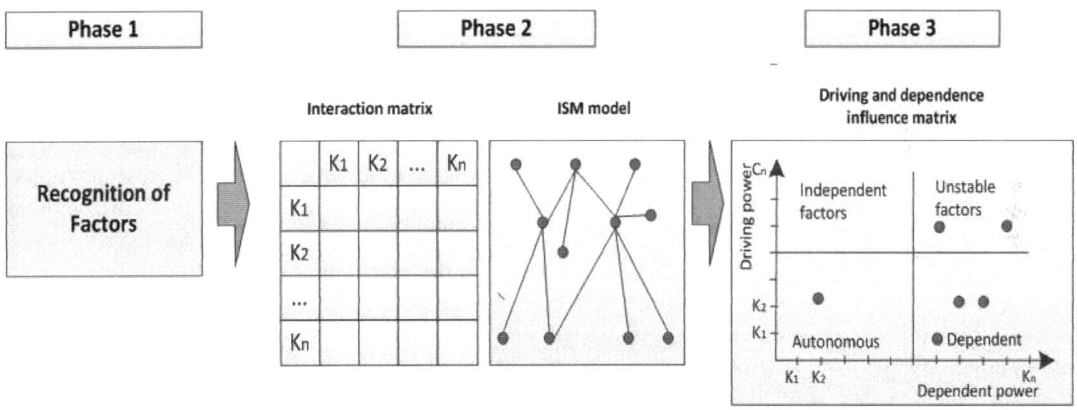

Figure 1. Simplified scheme of the procedure

In the Chapter 4, the MICMAC (Matrice d'Impacts crossovers-multiplication appliquée anclassment) analysis was conducted, which allowed for an assessment of the influence and dependence of the variables and the visualization of the results. A comparative analysis of the results obtained for a large water distribution network supporting urban agglomerations was also conducted with the results obtained for the water distribution network serving the small municipality. The last chapter is a summary and identification of directions for further research.

2. Interpretive Structural Modeling - ISM

In the literature, three main techniques like: Interpretive Structural Modeling, Analytic Network Process and Analytic Hierarchy Process are used for modeling and multi criteria decision making (MCDM) in the different area [7, 8, 9]. Jitesh Thakkar et al. [10] have compared these three techniques and extracts from that comparison is shown in the Table 1. This comparison aims to represent the main advantages of ISM technique.

Table 1. Brief comparison between AHP, ANP and ISM [10]

Analytic Hierarchy process (AHP)	Analytic Network Process(ANP)	Interpretive Structural modeling Technique (ISM)
Discipline of hierarchy has to be strictly followed	Deals with loose networks	Involves a set of Interconnected criteria
Fails in complex real life problems	Useful in real life non-linear problems	Captures the complexities of real life problems
Assumes functional independence of an upper part of hierarchy from its lower one	Takes into account the Interdependencies and non-linearity	Establishes the "leads to" relationships among the criteria
Moderate ability for capturing dynamic complexity	Lower ability for capturing complexity	Higher ability for capturing dynamic complexity

Interpretive Structural Modeling was first proposed by J. Warfield [11] to analyze the complex socioeconomic systems. ISM is a computer-assisted learning process that enables individuals or groups to develop a map of the complex relationships between the many elements involved in a complex situation. Its basic idea is to use experts' practical experience and knowledge to decompose a complicated system into several sub-systems (elements) and construct a multilevel structural model.

Warfield developed a methodology that uses systematic application of some elementary notions of graph theory and Boolean algebra in such a way that when implemented in a man machine interactive mode, theoretical, conceptual and computational leverage is exploited to construct directed graph (a representation of the hierarchical structure of the system). This methodology has at least two desirable properties when compared to the similar approaches namely simplicity in the sense of not requiring from the user i.e. viewpoint of advance mathematical knowledge and efficiency in terms of economizing in computer time.

The various steps involved in ISM modeling are as follows: (1) Identify the elements which are relevant to the problem. This could be done by a survey or group problem solving technique; (2) Establish a contextual relationship between elements with respect to which pairs of elements would be examined; (3) Develop a structural self-interaction matrix (SSIM) of elements. This matrix indicates the pair-wise relationship among elements of the system. This matrix is checked for transitivity; (4) Develop a reachability matrix from the SSIM; (5) Partition the reachability matrix into different levels; (6) Convert the reachability matrix into conical form; (7) Draw digraph based on the relationship given in reachability matrix and remove transitive links and convert the resultant digraph into an ISM based model by replacing element nodes with the statements; (8) Review the model to check for conceptual inconsistency and make the necessary modifications.

ISM is an interactive learning process. In this technique, a set of different directly and indirectly related elements are structured into a comprehensive systematic model. Applications of ISM in various systems/fields is shown in the Table 2.

Table 2. Applications of ISM in various systems/fields

Reference	System/field
Amrina et al., 2016 [12]	This paper aims to propose a fuzzy multi criteria approach for evaluating sustainable manufacturing in cement industry which integrated the Interpretive Structural Modeling (ISM) and the Fuzzy Analytic Network Process (FANP) methodology.
Kumar et. Al., 2013 [13]	The study developed a structural model of the variables, important to implement Lean Manufacturing System in Indian automobile industry. The model based upon experts' opinions. Eighteen variables have been identified from the literature and subsequent discussions with experts.
Panackal & Singh (2015) [14]	The study reflected various fields of contribution of rural youth towards sustainable development and significant important areas for improvement High interrelationship and interconnectivity between the rural youth and sustainability was derived from ISM Model.

	The analysis based on extensive literature review and ISM modelling resolved the contribution of youth in various ways towards rural sustainability
Mandal & Deshmukh (1994) [15]	Developed an analytical framework for the vendor selection process combining these qualitative and quantitative criteria. Vendor selection criteria have been identified from Dickson's study. Variables like production facilities and capacity technical capabilities, management and organisation and labour relation are the strong driver variables
Diabat & Govindan (2011) [16]	Developed a model of drivers affecting the implementation of green supply chain management using an interpretive structural modelling (ISM) framework. Eleven drivers that affect implementation of green supply chain were identified from the literature. The model can be validated using a SEM framework
Jadhav et al., (2014) [17]	Developed the complex patterns of contextual relationship in the development of framework for sustainable lean implementation using ISM. Two broad conceptual contributions are reported: First, it explored Lean practice bundles for successful implementation of Lean and second, it provided brief description of eight Lean practice bundles

3. Methodology

3.1. Introduction of the object of the research

The object of the research is a water distribution network supplying the Śmigiel municipality. The first plans for construction of the Śmigielski aqueduct were made in 1911. Currently the water distribution network extends over a total area of 185.5 km2, of which 134.3 km^2 is the rural area and 51.2 km^2 is urban area. The water distribution network consists of 6 water intake stations, and the total daily water production is 2146 m^3 and supplies about 18 thousand residents. Table 3 summarizes the length of the water distribution network of Śmigiel municipality in the years 2012 - 2016, while table 4 shows the age structure of the water distribution system elements.

Table 3. The lenght of the water distribution network in Śmigiel municipality in the years 2012-2016

	YEAR / Lenght of teh network [km]	2012	2013	2014	2015	2016
Type of the network	distribution	**187,7**	**187,7**	**189,2**	**189,2**	**189,2**

Table 4. Age structure of the water distribution network in Śmigiel municipality

Age of the ducts	Up to 5 years	between 6-10 years	between 11-20 years	over 20 years
Percentage	**5%**	**15%**	**20%**	**60%**

The network includes mainly pipes made of: cement-asbestos, PVC, cast iron and steel. The percentage of materials from which the water distribution network is made of is shown in Table 5

Table 5. Percentage of materials from which the water distribution network is made of in Śmigiel municipality

Material	PCV	Steel	Cast iron	PE	Cement-asbestos
Percentage	**65%**	**0%**	**15%**	**2%**	**18%**

The goal of the enterprise is to supply water to the consumers, providing: (1) quantitative parameters, taking into account needs of customers varying in time and space (2) quality parameters, taking into account the need to adjust and maintain the water characteristics at the level defined by legal requirements. In pursuit of such goals, the network's performance is crucial for the company, including:

- operation aspect - providing organizational and economic conditions for the efficient use of all network elements (sale of water produced to end users)
- maintenance aspect - providing technical capabilities for effective operation and efficient use of the network, by monitoring and maintaining parameters at the required level, e.g. water pressure in the network, carrying out scheduled inspections, maintenance and repairs and performing unscheduled work resulting from emergency events.

It is important for network managers to consider what external factors should be taken into account when determining the exploitation policy for water distribution network, and which of them will have the greatest impact both on the performance of the customer response and on the company's own efficiency

3.2. Identification of external factors influencing exploitation policy of Śmigiel water distribution system

The research was to identify, select and organize a set of features describing the impact of the external environment on the water distribution network's exploitation policy. Due to the objective of the research, the external environment features developed by A. Loska [5] for the large water distribution system were used for further consideration. The identification of the external environment was carried out in the research with an expert method (involving researchers and employees of the company) using the STEEPVL model. This model includes seven criterion classes, each of which determines the selected aspect of the external environment: Social, Technological, Economic, Ecological, Political, Value, Legal. The identified features for each of the criteria classes are shown in the Table 6.

Table 6. The list of the factors, as a result of the analysis using the STEEPVL model [5].

Criteria	Features (F)
Social	(1) Level of request and consumption of water in the activity area of the enterprise. (2) Staff potential of the enterprise. (3) Staff potential of the maintenance department of the enterprise
Technological	(4) Access to innovative exploitation solutions on the water supply system Level of modernity of the water supply system. (5) Access to innovative exploitation solutions on the water supply system (6) Level of the use of innovative methods and tools in the exploitation of the water supply
Economic	(7) Dynamics of changes in the price (cost) of water supply to consumers. (8) Expenditures for development and modernization of the water supply system. (9) Expenditures for activities of the maintenance department.
Ecological	(10) Influence of the water supply system on the environment. (11) Impact of exploitation activities for the water supply system on the environment.
Political	(12) State and region policy in the field of water management. (13) Enterprise policy for ensuring supply of water to customers. (14) Importance of the maintenance activity for the enterprise.
Value	(15) Life style and consumption patterns of the local community. (16) Level of acceptance of maintenance works of the water supply system by the local community.
Legal	(17) Legislation in the supply of water to consumers. (18) Legal requirements and standards for the exploitation of the water supply system.

The criteria and features presented in Table 6 were further investigated by a panel of experts

3.3. Interpretive Structural Modeling approach

Analysis of external factors influencing the definition of exploitation policy of the water distribution network in Śmigiel municipality was carried out with respect to the stages presented in figure 2. These stages were divided into five steps.

Step 1 Structural self-interaction matrix (SSIM)

For developing contextual relationships among variables, the opinion of total five experts, three from industry and two from academia have been taken. For expressing the relationship between different critical factors, four symbols have been used to denote the direction of relationship between the parameters i and j (here i , j): **V** is used for the relation from enabler i to enabler j (i.e. if enabler i influences or reaches to enabler j); **A** is used for the relation from enabler j to enabler i (i.e. if enabler j reaches to enabler i); **X** is used for both direction relations (i.e. if enablers i and j reach to each other); **O** is used for no relation between two enablers (i.e. if enablers i and j are unrelated). Based on this contextual relationship, the SSIM has been developed. To obtain consensus, the SSIM was discussed in a group of factory experts and based on their responses, SSIM has been finalized and is presented in table 7.

Table 7. Structural self-interaction matrix (SSIM)

F_i	1	2	3	4	5	6	7	8	9	10	11	12	13	14	15	16	17	18
1		V	V	0	0	V	V	0	0	0	0	X	V	V	A	V	0	V
2			X	0	0	A	A	X	0	0	0	0	A	A	0	0	0	0
3				0	0	X	0	0	V	0	0	0	A	X	0	X	0	0
4					A	A	0	X	0	V	0	0	A	V	0	V	0	0
5						V	0	0	A	0	0	0	V	A	0	0	0	V
6							V	X	X	0	X	0	0	A	0	X	0	A
7								A	A	A	A	A	A	0	X	V	A	0
8									V	X	V	A	X	X	0	0	A	A
9										V	X	0	A	X	0	A	0	A
10											V	0	0	0	0	0	A	0
11												0	0	1	0	0	0	0
12													V	0	0	0	V	0
13														V	A	0	0	A
14															0	X	0	A
15																V	A	0
16																	0	0
17																		0

Step 2 Reachability matrix

The SSIM is transformed into a reachability matrix format by transforming the information in each entry of the SSIM into 1s and 0s in the reachability matrix. The substitution of 1s and 0s are as per the following rules: (1) If the (i, j) entry in the SSIM is V, the (i, j) entry in the reachability matrix becomes 1 and the (j, i) entry becomes 0; (2) If the (i, j) entry in the SSIM is A, the (i, j) entry in the reachability matrix becomes 0 and the (j, i) entry becomes 1; (3) If the (i, j) entry in the SSIM is X, the (i, j) entry in the reachability matrix becomes 1 and the (j, i) entry also becomes 1; (4) If the (i, j) entry in the SSIM is O, the (i, j) entry in the reachability matrix becomes 0 and the (j, i) entry also becomes 0. Following the above rules, the initial reachability matrix is prepared and is shown in tab.8.

Table 8. Initial reachability matrix

F_i	1	2	3	4	5	6	7	8	9	10	11	12	13	14	15	16	17	18
1	0	1	1	0	0	1	1	0	0	0	0	1	1	1	0	1	0	1
2	0	0	1	0	0	0	0	1	0	0	0	0	0	0	0	0	0	0
3	0	1	0	0	0	1	0	0	1	0	0	0	0	1	0	1	0	0
4	0	0	0	0	0	0	0	1	0	1	0	0	0	1	0	1	0	0
5	0	0	0	1	0	1	0	0	0	0	0	0	1	0	0	0	0	1
6	0	1	1	1	0	0	1	1	1	0	1	0	0	0	0	1	0	0
7	0	1	0	0	0	0	0	0	0	0	0	0	0	0	1	1	0	0
8	0	1	0	1	0	1	1	0	1	1	1	0	1	1	0	0	0	0
9	0	0	0	0	1	1	1	0	0	1	1	0	0	1	0	0	0	0
10	0	0	0	0	0	0	1	1	0	0	1	0	0	0	0	0	0	0
11	0	0	0	0	0	1	1	0	1	0	0	0	0	1	0	0	0	0
12	1	0	0	0	0	0	1	1	0	0	0	0	1	0	0	0	1	0
13	0	1	1	1	0	0	1	1	1	0	0	0	0	1	0	0	0	0
14	0	1	1	0	1	1	0	1	1	0	1	0	0	0	0	1	0	0
15	1	0	0	0	0	0	1	0	0	0	0	0	1	0	0	1	0	0
16	0	0	1	0	0	1	0	0	1	0	0	0	0	1	0	0	0	0
17	0	0	0	0	0	0	1	1	0	1	0	0	0	0	1	0	0	0
18	0	0	0	0	0	1	0	1	1	0	1	0	1	1	0	0	0	0

1* entries are included to incorporate transitivity to fill the gap, if any, in the opinion collected during development of structural self-instructional matrix. After incorporating the transitivity concept as described above, the final reachability matrix is obtained (table 9).

Table 9. Final reachability matrix

F_i	1	2	3	4	5	6	7	8	9	10	11	12	13	14	15	16	17	18
1	1	1	1	1*	1*	1	1	1*	1*	1*	1*	1	1	1	1*	1	1*	1
2	0	1	1	1*	0	1*	1*	1	1*	1*	1*	0	1*	1*	0	1*	0	0
3	0	1	1	1*	1*	1	1*	1*	1	1*	1*	0	0	1	0	1	0	0
4	0	1*	1*	1	1*	1*	1*	1	1*	1	1*	1*	1*	1	0	1	0	0
5	0	1*	1*	1	1	1	1*	1*	1*	1*	1*	0	1	1*	0	1*	0	1
6	0	1	1	1	0	1	1	1	1	1*	1	0	1*	1*	1*	1	0	0
7	1*	1	1*	0	0	1*	1	1*	1*	0	0	0	1*	1*	1	1	0	0
8	0	1	1*	1	0	1	1	1	1	1	1	0	1	1	0	1*	0	0
9	0	1*	1*	1*	1	1	1	1*	1	1	1	0	1*	1	1*	1*	0	1*
10	0	1*	0	1*	0	1*	1	1	1*	1	1	0	1*	1*	1*	1*	0	0
11	0	1*	1*	1*	1*	1	1	1*	1	1*	1	0	0	1	1*	1*	0	0
12	1	1*	1*	1*	0	1*	1	1	1*	1*	1*	1	1	1*	1*	1*	1	1*
13	0	1	1	1	0	1*	1	1	1	0	0	0	1	1	0	1*	0	0
14	0	1	1	1*	1	1	1*	1	1	1*	1	0	1*	1	0	1	0	1*
15	1	1*	1*	0	0	1*	1	1*	1*	0	0	1*	1	1*	1	1	0	1*
16	0	1*	1	1*	1*	1	1*	1*	1	1*	1*	0	0	1	0	1	0	0
17	0	1*	0	0	0	0	1	1	1*	1	1*	0	1*	1*	1	1*	1	0
18	0	1*	1*	1*	0	1	1*	1	1	1*	1	0	1	1	1*	1*	0	1

Step 3 Partitioning the reachability matrix

The matrix is partitioned, by assessing the reachability and antecedent sets for each variable. The reachability set consists of the element itself and other elements, which it may help to achieve, whereas the antecedent set consists of the element itself and other elements, which may help achieving it. Thereafter the intersection of these sets is derived for all the elements. The elements for which the reachability and the intersection sets are the same occupy the top level in the ISM hierarchy. The top-level element in the hierarchy would not help achieve any other element above its own level. Once the top-level element is identified, it is separated out from the other elements. Then, the same process is repeated to find out the elements in the next level. This process is continued until the level of each element is found. These levels help in building the diagraph and the ISM model. In the present case, the 18 enablers, along with their reachability set, antecedent set, intersection set and levels, are presented in table 10.

Table 10. Reachability, antecedent and intersection sets for reachability matrix

F_i	Reachability set $R(p_i)$	Antecedent set $A(p_i)$	Intersection set $R(p_i) \cap A(p_i)$	Level
1	1, 2,3,4,5,6,7,8,9,10,11,12,13, 14,15,16,17,18	1,7,12,15	1,7,12,15	X
2	2,3,4,6,7,8, 9,10,11, 13,14,16	1,2,3,4,5,6,7,8,9,10,11,12,13 ,14, 15,16,17,18	2,3,4,6,7,8, 9,10,11, 13,14,16	III
3	2,3,4,5,6,7,8, 9,10,11,14,16	1,2,3,4,5,6,7,8,9,11,12,13,14 ,15, 16,18	2,3,4,5,6,7,8, 9,10,11,14,16	IV
4	2,3,4,5,6,7,8, 9,10,11,12,13,14,16	1,2,3,4,5,6, 8,9,10,11,12,13,14, 16,18	2,3,4,5,6,7,8, 9,10,11,12,13,14,16	V
5	2,3,4,5,6,7,8, 9,10,11,13,14,16,18	1,3,4,5,9,11, 14,16	3,4,5,9,11, 14,16	IX
6	2,3,4,6,7,8,9,10,11,13,14,15,16	1,2,3,4,5,6,7,8,9,10,11,12,13 ,14, 15,16,18	2,3,4,6,7,8,9,10,11, 13,14,15,16	IV
7	1,2,3,6,7,8,9,13,14,15,16	1,2,3,4,5,6,7,8,9,10,11,12,13 ,14, 15,16, 17,18	1,2,3,6,7,8,9,13,14, 15,16	II
8	2,3,4,6,7,8,9,10,11,13,14,16	1,2,3,4,5,6,7,8,9,10,11,12,13 ,14, 15,16, 17,18	2,3,4,6,7,8,9,10,11, 13,14,16	III
9	2,3,4,5,6,7,8,9,10,11,13,14,15,16	1,2,3,4,5,6,7,8,9,10,11,12,13 ,14, 15,16,17,18	2,3,4,5,6,7,8,9,10,1 1, 13,14,15,16,	V
10	2,4,6,7,8,9,10,11,13,14,15,16	1, 2, 3, 4,5,6, 8,9,10,11,12, 14, 16,17,18	2,4,6, 8,9,10,11, 14, 16	III
11	2,3,4,5,6,7,8,9,10,11,14,15,16	1, 2, 3, 4,5,6, 8,9,10,11,12, 14, 16,17,18	2,3,4,5,6, 8,9,10,11,14,16	VI I
12	1,2,3,4,6,7,8,9,10,11,12,13,14,15, 16,1718	1,4,12,15	1,4,12,15	X
13	2,3,4,6,7,8,9,13,14,16	1, 2, 4,5,6,7,8,9,10, 12,13,14,15, 17,18	2, 4,6,7,8,9,13,14	I
14	2,3,4,5,6,7,8,9,10,11,13,14, 16,18	1,2,3,4,5,6,7,8,9,10,11,12,13 ,14, 15,16,17,18	2,3,4,5,6,7,8,9,10,1 1, 13,14, 16,18	V
15	1,2,3,6,7,8,9,12,13,14,15,16,18	1,6,7,9,10,11,12,15,17,18	1,6,7,9,12,15,18	VI
16	2,3,4,5,6,7,8,9,10,11,14,16	1,2,3,4,5,6,7,8,9,10,11,12,13 ,14, 15,16,17,18	2,3,4,5,6,7,8,9,10,1 1, 13,14, 16,18	IV
17	2,7,8,9,10,11,13,14,15,16,17	1,12,17	17	VI II
18	2,3,4,6,7,8,9,10,11,13,14,15,16,18	1,5,9,12,14,15,18	9,14,18	VI II

Step 4 Development of conical matrix

Conical matrix is developed by clustering factors in the same level across the rows and columns of the final reachability matrix. The drive power (DRP) of a factor is derived by summing up the number of ones in the rows and its dependence power (DEP) by summing up the number of ones in the columns. Next, drive power and dependence power ranks are calculated by giving highest ranks to the factors that have the maximum number of ones in the rows and columns, respectively (Table 11).

Table 11. Conical matrix

F_i	1	2	3	4	5	6	7	8	9	10	11	12	13	14	15	16	17	18	DRP
1	1	1	1	1*	1*	1	1	1*	1*	1*	1*	1	1	1	1*	1	1*	1	18
2	0	1	1	1*	0	1*	1*	1	1*	1*	1*	0	1*	1*	0	1*	0	0	12
3	0	1	1	1*	1*	1	1*	1*	1	1*	1*	0	0	1	0	1	0	0	12
4	0	1*	1*	1	1*	1*	1*	1	1*	1	1*	1*	1*	1	0	1	0	0	14
5	0	1*	1*	1	1	1	1*	1*	1*	1*	1*	0	1	1*	0	1*	0	1	14
6	0	1	1	1	0	1	1	1	1	1*	1	0	1*	1*	1*	1	0	0	13
7	1*	1	1*	0	0	1*	1	1*	1*	0	0	0	1*	1*	1	1	0	0	11
8	0	1	1*	1	0	1	1	1	1	1	1	0	1	1	0	1*	0	0	12
9	0	1*	1*	1*	1	1	1	1*	1	1	1	0	1*	1	1*	1*	0	1*	15
10	0	1*	0	1*	0	1*	1	1	1*	1	1	0	1*	1*	1*	1*	0	0	12
11	0	1*	1*	1*	1*	1	1	1*	1	1*	1	0	0	1	1*	1*	0	0	13
12	1	1*	1*	1*	0	1*	1	1	1*	1*	1*	1	1	1*	1*	1*	1	1*	17
13	0	1	1	1	0	1*	1	1	1	0	0	0	1	1	0	1*	0	0	10
14	0	1	1	1*	1	1	1*	1	1	1*	1	0	1*	1	0	1	0	1*	14
15	1	1*	1*	0	0	1*	1	1*	1*	0	0	1*	1	1*	1	1	0	1*	13
16	0	1*	1	1*	1*	1	1*	1*	1	1*	1*	0	0	1	0	1	0	0	12
17	0	1*	0	0	0	0	1	1	1*	1	1*	0	1*	1*	1	1*	1	0	11
18	0	1*	1*	1*	0	1	1*	1	1	1*	1	0	1	1	1*	1*	0	1	14
DEP	4	18	16	15	8	17	18	18	18	15	15	4	15	18	9	18	3	7	

Step 5 Development of digraph and ISM model

From the conical form of reachability matrix, the preliminary digraph including transitive links is obtained. It is generated by nodes and lines of edges. After removing the indirect links, a final digraph is developed. In this development, the top level factor is positioned at the top of the digraph and second level factor is placed at second position and so on, until the bottom level is placed at the lowest position in the digraph.

4. MICMAC analysis

Matrice d'Impacts croises-multiplication appliqu′e anclassment (cross-impact matrix multiplication applied to classification) is abbreviated as MICMAC. The purpose of MICMAC analysis is to analyze the drive power and dependence power of enablers. MICMAC principle is based on multiplication properties of matrices. It is done to identify the key enablers that drive the system in various categories. Based on their drive power and dependence power, the enablers, have been classified into four categories as follows: (1) Autonomous enablers: These enablers have weak drive power and weak dependence power. They are relatively disconnected from the system, with which they have few links, which may be very strong; (2) Linkage enablers: These enablers have strong drive power as well as strong dependence power. They enablers are unstable in the fact that any action on these enablers will have a effect on others and also a feedback effect on themselves; (3) Dependent enablers: These enablers have weak drive power but strong dependence power (4) Independent enablers: These enablers have strong drive power but weak dependence power. An enabler with a very strong drive power, called the 'key enabler' falls into the category of independent or linkage enablers. The drive power and dependence power of enablers is shown in table 11. The drive power - dependence power diagram is drawn as shown in Fig. 2 This Figure has been divided into four clusters. First cluster includes 'autonomous enablers', second cluster includes 'dependent enablers', third cluster includes 'linkage enablers' and fourth cluster contains 'independent enablers'.

There are for example (17) Legislation in the supply of water to consumers; (15) Life style and consumption patterns of the local community factors in the autonomous cluster. The next cluster consists of dependent variables like: (10) Influence of the water supply system on the environment, (3) Staff potential of the maintenance department of the enterprise, (11) Impact of exploitation activities for the water supply system on the environment. In this particular cluster, these variables have the least driving power and have highest dependence and form the topmost level in the ISM hierarchy. The managers should take special care for handling these enablers. The next cluster consists of those variables that are termed as linkage variables like (9) Expenditures for activities of the maintenance department and (4) Access to innovative exploitation solutions on the water supply system Level of modernity of the water supply system, which are influenced by lower level variables and in turn impacts other variables in the ISM model. The last cluster includes independent variables like (1) Level of request and consumption of water in the activity area of the enterprise and (12) State and region policy in the field of water management. These variables have strong driving power and weak dependency on other enablers. They may be treated as the 'key enablers' for the successful implementation of exploitation policy for water distribution network.

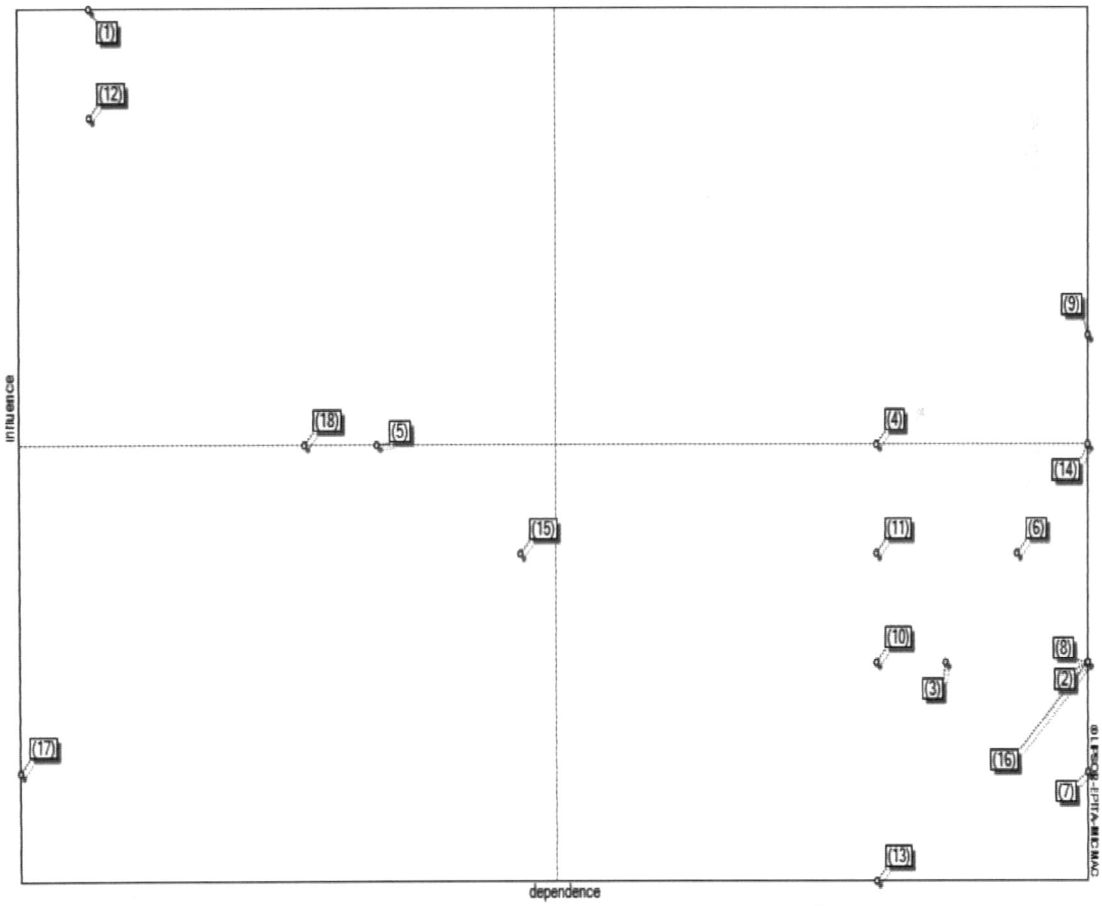

Figure 2. MICMAC Direct influence/dependence map (Śmigiel)

To answer the question: Does the size and nature of the water distribution system significantly affect the adopted exploitation policy? It is necessary to compare the research results for the large metropolitan water distribution system (MS) with the results obtained in the research presented in the paper. Because the method described by J. Nazarko [18] was used in the research (MS), the results were to be transposed into the ISM matrix so that the data prepared could be used to develop the MICMAG graph. Graphs of water distribution networks compared are shown in the Figure 3.

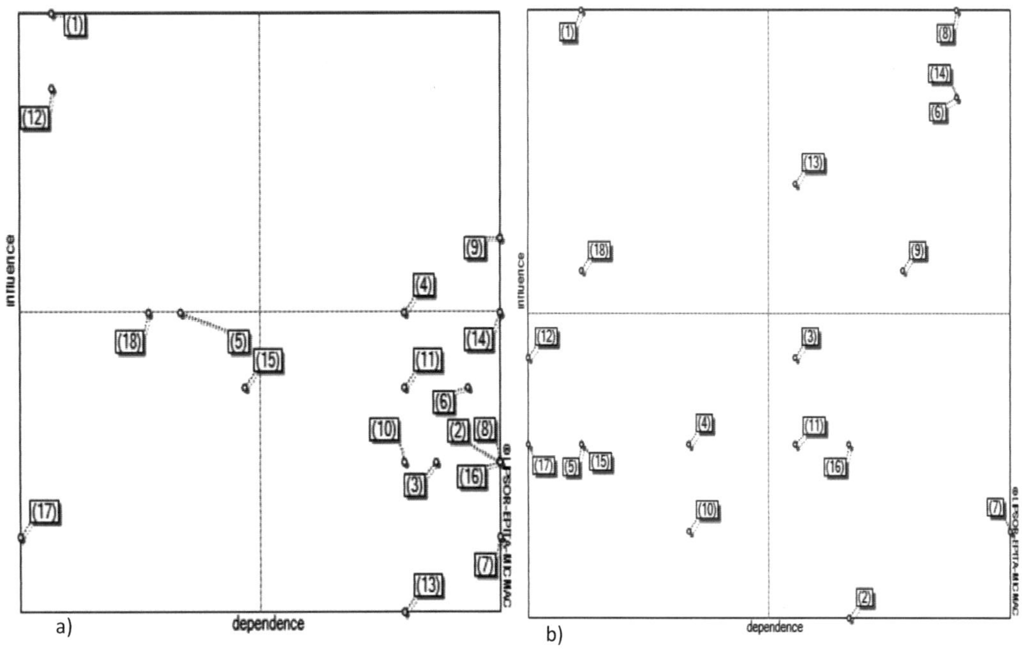

Figure 3. Comparison of MICMAC direct influence/dependence map: a) for described small WDS, b) for WDS described in [6]

The comparison showed that depending on the size of the water distribution system, different criteria are of various importance for the water distribution network exploitation policy definition. There is considerable variation in the approach to maintenance and repair (14), network development and modernization (8) and the use of innovative solutions (6). In a large enterprise, these factors are considered as the crucial ones - large enterprises have the resources and capabilities to innovate and plan for the long-term exploitation of the network, shaping its future prospects. On the one hand, it seems advisable to pursue further action striving for identifying the impact of individual factors on corporate policy and identifying undisclosed factors, on the other hand it signals the need for search for the reasons why innovation is less important as innovation - Is it just a result of smaller financial resources, or are there other causes? Criteria that in both cases are of comparable importance, are, among others (1) (9) and (17) - companies are striving to achieve their primary goal, i.e., water distribution (probably) minimizing maintenance expenditure, paying attention to legal regulations in this regard but treating them as external factor, which they have no influence on.

5. Conclusion

The major objective of this paper is to identify the factors that significantly affect the exploitation policy for water distribution system. In this paper, an ISM-based model has been developed to analyze the interactions among different external factors influencing water distribution system in a small municipality.

The ISM model developed in this paper acts as a tool for top management to understand/identify the key enablers of external factors influencing exploitation policy for water distribution system in a small municipality. This model has been developed on the basis of consensus of experts from industry and authors of this paper. Comparison between two water distribution network (small and large metropolitan water distribution system) indicate that 3 factors are of comparable importance, are, among others (1) (9) and (17) - companies are striving to achieve their primary goal, i.e., water distribution (probably) minimizing maintenance expenditure, paying attention to legal regulations in this regard but treating them as external factor, which they have no influence on

References

[1] I Rojek 2010 Hybrid Neural Networks as Prediction Models. In: Rutkowski L , Scherer R , Tadeusiewicz R , Zadeh L A and Zurada J M. (eds) *Artifical Intelligence and Soft Computing. ICAISC 2010. Lecture Notes in Computer Science* vol **6114** pp 88-95

[2] I Rojek 2014 *Models for Better Environmental Intelligent Management within Water Supply Systems Water Resour Manage* vol **28(12)** pp 3875-90

[3] E Dostatni, J Diakun, D Grajewski, R Wichniarek and A Karwasz 2015 *Multi-agent System to Support Decision-Making Process in Ecodesign. In: Herrero et. al., (eds) 10th International Conference on Soft Computing Models in Industrial and Environmental Applications. Advances in Intelligent Systems and Computing* vol **368** pp 463-74

[4] A Karwasz, E Dostatni, J Diakun, D Grajewski, R Wichniarek, M Stachura 2016 *Estimating the cost of product recycling with the use of ecodesign support system, management and production engineering review* Vol **7(1)** pp 33-39

[5] A Loska 2017 *Scenario modeling exploitation decision-making process in technical network systems. Eksploatacja i Niezawodność – Maintenance and Reliability* **19 (2)** pp 268–78

[6] A Loska 2016 *Methodology of modelling exploitation decision-making process for technical netwosk systems (in polish) WPS Gliwice*

[7] J Baran, A Janik, A Ryszko and M Szafraniec 2016 *Selected environmental methods and tools supporting eco-innovation implementation within national smart specialisations in Poland, 3rd International Multidisciplinary Scientific Conference on Social Sciences and Arts SGEM 2016, SGEM2016 Conference Proceedings, Book 2* Vol **3** pp1029-36 Albena, Bulgaria

[8] Kujawińska A and Rogalewicz M 2017 Adaptive methods of process state evaluation – the development of an application for engineering purposes. In: Abraham A *et al., (eds.), Proceedings of the Third International Afro-European Conference for Industrial Advancement — AECIA 2016, Advances in Intelligent Systems and Computing, Springer* **565** pp 47-56

[9] A Burduk 2015 The role of artificial neural network models in ensuring the stability of systems, 10th International Conference on Soft Computing Models in Industrial and Environmental Applications In: *Ávaro Herrero (eds.) Springer* **vol. 368** pp 427-37

[10] J Thakkar, A Kanda and S G Deshmukh 2008 *Interpretive structural modeling (ISM) of IT-enablers for Indian manufacturing SMEs, Information Management & Computer Security* Vol. **16(2)** pp 13-136

[11] J Warfield 1974 *Developing interconnected matrices in structural modeling. IEEE Transaction Systems Man and Cybernetics* **4(1)** pp 51–81

[12] E Amrina, C Ramadhani, A L Vilsi and A Fuzzy 2016 *Multi Criteria Approach for Sustainable Manufacturing Evaluation in Cement Industry Procedia CIRP* **40** pp 619 –24

[13] Kumar N, Kumar S, Haleem A, Gahlot P 2013 Implementing Lean Manufacturing System: *ISM approach. Journal of Industrial Engineering and Management* **6(4)** pp 996-1012

[14] N Panackal, A Singh 2015 *Using Interpretive Structural Modeling to Determine the Relation between Youth and Sustainable Rural Development. IBMRD's Journal of Management & Research* **4 (1)** pp 58–74

[15] A Mandal, S G Deshmukh 1994 *Vendor Selection Using Interpretive Structural Modelling (ISM). International Journal of Operations & Production Management* **14 (6)** pp 52–59

[16] A Diabat and K. Govindan 2011 *An Analysis of the Drivers Affecting the Implementation of Green Supply Chain Management. Resources, Conservation and Recycling* **55 (6)** pp 659–667

[17] J R Jadhav, S S Mantha, S B Rane 2014 *Development of Framework for Sustainable Lean Implementation: An ISM Approach. Journal of Industrial Engineering International* **10 (3)** pp 1–27

[18] J Nazarko and K A Kuźmicz 2017 *Introduction to the STEEPVL Analysis of the New Silk Road Initiative, Procedia Engineering* Vol **182** pp 497-503

Measurements of K X-ray fluorescence cross-sections, fluorescence yields, level widths and radiative vacancy transition probabilities for the elements Zr, Mo, Cd, Er at 59.54 keV

Mehmet Fatih Turhan[1], Ferdi Akman[2,*], Mustafa Recep Kaçal[3] and Rıdvan Durak[4]

[1]Afyon Kocatepe University, Ataturk Vocational School of Health Services, Medical Imaging Techniques, 03200 Afyonkarahisar, Turkey.
[2]Bingol University, Vocational School of Technical Sciences, Department of ElectronicCommunication Technology, 12000 Bingol, Turkey.
[3]Giresun University, Arts and Sciences Faculty, Department of Physics, 28100 Giresun, Turkey.Afyon Kocatepe
[4]Ataturk University, Sciences Faculty, Department of Physics, 25240 Erzurum, Turkey.

[*]Corresponding Author: fakman@bingol.edu.tr

Abstract. K shell X ray fluorescence cross sections, fluorescence yields, level widths and radiative vacancy transition probabilities were investigated for the elements Zr, Mo, Cd, Er. The measurements were performed using photons of 59.54 keV emitted from an Am-241 (about 100 mCi) point source and a high resolution Si(Li) detector. The experimental results were compared with theoretical calculations and the other results obtained from literature.

1. Introduction

One of the techniques used worldwide in many fields of science is energy dispersive X-ray fluorescence technique because EDXRF is comparatively low-cost and requires less technical effort to run the system. This technique is very useful for determining physical parameters such as fluorescence cross sections, fluorescence yields, level widths and vacancy transfer probabilities, intensity ratios, etc. The knowledge obtained from these physical parameters for different elements, compounds, polymers, thin films, ores, food productions, ceramics is very important in the study of some basic phenomena such as atomic, molecular and radiation physics, medical physics, environmental science, agriculture, nuclear science, trace element analysis, etc.

Several investigators determined the X-ray fluorescence cross sections, fluorescence yields, level widths, intensity ratios [1-19]. The semi-empirically fitted values of K shell X-ray fluorescence, auger and Coster–Kronig transition probabilities for all elements in the atomic number range 5≤Z≤110 were determined by Krause, [1]. K and L X ray fluorescence cross sections and fluorescence yields for some elements in the atomic range 22≤Z≤50 at 59,54 keV incidents photon were investigated by Yilmaz et al., [2], Şimşek et al. [3], Durak et al. [4], Han et al. [5], Akman et al. [6]. Gudennavar et al. [7], Durak [8] measured K-shell fluorescence yield, fluorescence cross-section and ratio between the radiative transition width and the Auger transition width for some elements in the atomic range from 30≤Z≤70 at 123,6 keV incidents photon. Krause and Oliver [9] presented the semi-empirical values of the natural widths of the K, L_1, L_2 and L_3 levels, $K_{\alpha1}$, $K_{\alpha2}$ X-ray lines and KL_1L_1, KL_1L_2 and KL_2L_3 auger lines for the elements 10≤Z≤110. Campbell and Papp [10] combined atomic level widths from experimental measurements and they calculated theoretical widths based upon the Dirac-Hartree-Slater version of the independent particle model. K shell intensity ratios, production cross sections, fluorescence yields, vacancy transfer probabilities from K to L shell, level widths and line widths for some heavy elements in the atomic range 73≤Z≤81 were measured by Cengiz et al. [11]. Ertugrul [12] measured fluorescence yields and level widths of some atomic sub-shells for Th and U using measurements of the L X-ray from targets. Ertuğrul [13] evaluated K shell radiative transition probabilities and K, L_2 and L_3 shell/subshell fluorescence yields for Zn, As, Se, Rb, Sr, Y and Zr.The K to Li (i=2, 3), K to L, and K to M shell vacancy transfer probabilities for La, Ce, Pr, Nd, Sm, Eu, Gd, Tb, Dy and Er were determined at 59.54 keV using a reflection geometry by Akman [14].Bennal et al. [15] determined the K-L total vacancy transfer probabilities (η_{KL}) of selected elements in the atomic range 42≤Z≤82 using a weak gamma source.Using the intensity ratios, the radiative vacancy transfer probabilities from the shell K to the L sub-shells for selected rare earth were determined by Herrera and Miranda [16].Bonzi [17] investigated the the radiative vacancy transfer probabilities from L_3 to M shell and L_3 to N shell for W, Re and Pb. Vacancy transfer probabilities from L_3 sub-shell to M, N and O shells and sub-shells for the elements Hf, Ta, W, Re, Pt, Au, HgO, Tl, Pb, Bi, Th, U were measured using L shell fluorescence yields and X-ray intensity ratios by Tuzluca et al. [18]. Turhan et al. [19] investigated L X-ray fluorescence cross sections, L sub-shell fluorescence yields, level widths and radiative vacancy transfer probabilities of L sub-shells to Mi, Ni and Oi sub-shells for the elements Ho, Lu, W, Hg and Bi.

In this study, we determined K shell fluorescence cross sections, fluorescence yields, level widths and K to L_2, L_3, M_2, M_3 radiative vacancy transfer probabilities for Zr, Mo, Cd, Er. The determined experimental parameters are compared with theoretical values and the other available experimental values. The results in the present paper are found to be in good agreement with theoretical values and the other results obtained from literature.

2. Experimental Procedure

The experimental arrangement is shown in Fig. 1.

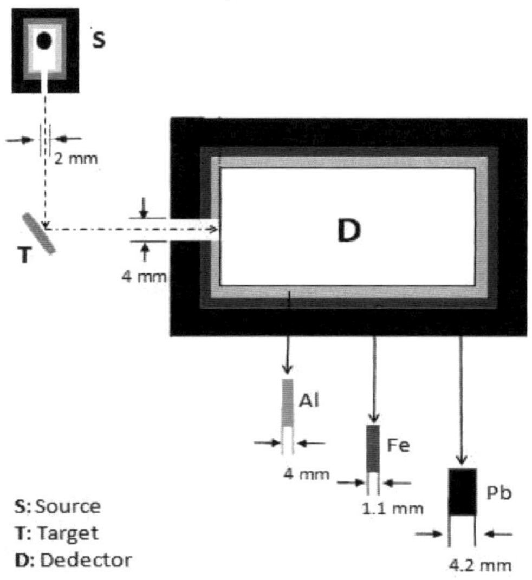

Figure 1. Experimental set-up.

The samples were excited using photons of 59.54 keV emitted from Am-241 (about 100 mCi) point source. Spectroscopically pure circular disc (diameter=13 mm) targets of Zr, Mo, Cd and Er of thicknesses ranging from 0.26 to 0.33 g/cm^2 have been used for measurements. The K X-rays were detected by a Si(Li) detector (active area 12 mm^2, sensitive depth 3 mm, thickness of Be window 0.025 mm and energy resolution of 185 eV at 5.9 keV). The detector was also placed in a step-down shield made from Pb, Fe and Al to minimize the detection of any radiation coming directly from the source and scattered from the surroundings. The spectra were accumulated in time intervals from 3600 to 14400 s in order to obtain sufficient statistical accuracy. In this study, peak integrals were analyzed by using Microcal Origin 7.0 software demo program with least squares fit method. The net peak areas were separated by fitting the measured spectra with multi-Gaussian function plus polynomial backgrounds. Fig. 2 shows a typical K X-rays spectrum for the Er target at 59.54 keV excitation energy.

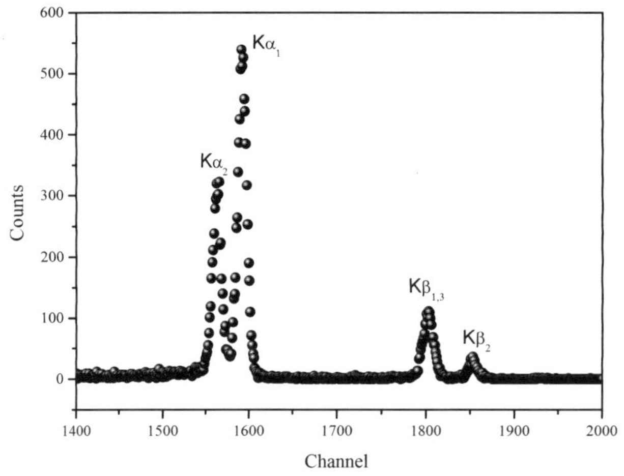

Figure 2. Typical K X-ray spectrum for Er at 59.54 keV.

3. Experimental and Calculation Procedures

The values of experimental K X-ray fluorescence cross-sections were determined using the following equation:

$$\sigma_{Ki} = \frac{N_{Ki}}{I_0 G \varepsilon_{Ki} \beta_{Ki} t} (i=\alpha, \beta) \quad (1)$$

where N_{Ki} is the number of counts per second under the corresponding K_iX-ray photopeak,I_0 is the intensity of the exciting radiation, G is the geometrical factor dependent on the source-sample-detector geometry, ε_{Ki} is the detector efficiency for K_iX-rays, t is the thickness of target in g/cm^2 unit and β_{Ki} is the self-absorption correction factor for the target material and is given by:

$$\beta_{Ki} = \frac{1-\exp\left[-\left(\mu_i/\cos\theta_i + \mu_e/\cos\theta_e\right)t\right]}{\left(\mu_i/\cos\theta_i + \mu_e/\cos\theta_e\right)t} \quad (2)$$

where μ_i and μ_e are the total mass absorption coefficients of the target material at the incident photon (59.54 keV) and emitted K_i X-ray energies, respectively. The values of μ_i and μ_e were taken from WinXCom program [20] being a Windows version of XCOM[21] on the basis of mixture rule. The angles of Θ_i and Θ_e which correspond incident and emission angles with sample surface were set to 45^0, respectively.

The detector efficiency $I_0G\varepsilon_{Ki}$, which contain terms related to the incident photon flux, geometrical factor and the efficiency of the X-ray detector was determined by measuring t; β_{Ki} and the K X-ray intensities from samples of Ti, Cr, Ni, Ru, Ag, Sn, Ce, Pr, Gd, Tb and Ho and using theoretical σ_{Ki} values:

$$I_0 G \varepsilon_{Ki} = \frac{N_{Ki}}{\sigma_{Ki} \beta_{Ki} t} \qquad (i=\alpha, \beta) \quad (3)$$

where N_{Ki} is the number of counts per second under the corresponding K_i X-ray photopeak, $\sigma_{Ki}$is the fluorescence cross sections.

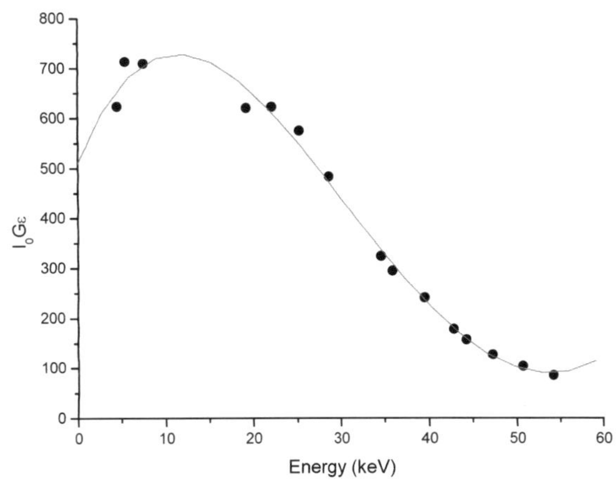

Figure 3. $I_0G\varepsilon$ values as a function mean *K* X-ray energy (*keV*).

The theoretical values of σ_{Ki}fluorescence cross sections were calculated using the following equation:

$$\sigma_{Ki} = \sigma_K(E)\omega_K F_{Ki} \tag{4}$$

where $\sigma_K(E)$ is the K-shell photoionization cross section taken from Scofield [22] for the elements at the excitation energy, ω_K is the K-shell fluorescence yield taken from Krause [1] and F_{Ki} is the fractional ratio of the K_i X-rays and $F_{K\alpha}$ and $F_{K\beta}$ are defined as:

$$F_{K\alpha} = \left(1 + \frac{I_{K\beta}}{I_{K\alpha}}\right)^{-1} \tag{5}$$

$$F_{K\beta} = \left(1 + \frac{I_{K\alpha}}{I_{K\beta}}\right)^{-1} \tag{6}$$

where $I_{K\alpha}$ and $I_{K\beta}$ are the $K\alpha$ and $K\beta$X-ray intensities, respectively. The theoretical values of $I_{K\beta}/I_{K\alpha}$were taken from Scofield [23].The measured $I_0G\varepsilon_{Ki}$ values for the present geometry were plotted as a function of the average K X-ray energy in Fig. 3.

The experimental Kshell fluorescence yields were obtained from using the following expression:

$$\omega_K = \frac{\sum \sigma_{Ki}}{\sigma_K(E)} (i=\alpha,\ \beta) \tag{7}$$

where $\sum \sigma_{Ki}$ is the total K X-ray fluorescence cross-section obtained experimentally, $\sigma_K(E)$is the theoretical K-shell photoionization cross-section of a given element for the excitation energy taken from Scofield [22].

The experimental K shell level widths were evaluated from the following equation:

$$\Gamma_K = \frac{\Gamma_K(R)}{\omega_K} \tag{8}$$

where $\Gamma_K(R)$ is radiative transition rate (Scofield, [23]), ω_K is the K shell fluorescence yield obtained experimentally.

Experimental radiative vacancy transition probabilities from K shell to L_2, L_3, M_2 and M_3 sub-shell were investigated from following equation:

$$\eta_{XiYt}(R) = \frac{\Gamma_{XiYt}(R)}{\Gamma_K} \tag{9}$$

Γ_{XiYt} (R)is the partial radiative transition rate and taken from Scofield [23], Γ_K is the K shell level width obtained experimentally.

4. Result and Discussion

K shell fluorescence cross sections from eq. 1 for Zr, Mo, Cd, Er at 59,54 keV excitation energy are listed in Table 1 along with the theoretical calculations (eq. 4) by using photoionization cross section (Scofield, [22]), fluorescence yield (Krause, [1]) and fractional emission rates in eqs. (5) and (6). The unit of all cross sections is barns/atom. It is seen from Table 1 that the experimental values are generally agreement with theoretical predictions within experimental errors (the range < 5.7%, < 16.5%, < 4.9%, respectively). The errors are attributed to uncertainties in different parameters used to evaluate K X-ray fluorescence cross sections, namely, the evaluation of peak areas (≤ 4.6%), $I_0G\varepsilon$(≤ 5.4%), thickness (≈ 2%) and absorption correction factor (≈ 2%).

Table 1. Experimental and theoretical K_i (i= α, β, total) X-ray fluorescence cross sections (barns/atom) values.

	$\sigma_{K\alpha}$		$\sigma_{K\beta}$		σ_{Ktotal}	
Element	**Experimental**	**Theoretical**	**Experimental**	**Theoretical**	**Experimental**	**Theoretical**
^{40}Zr	293.89±16.52	291.22	57.08±3.28	50.58	350.96±27.89	341.80
^{42}Mo	371.90±20.81	365.37	74.41±4.27	66.10	446.31±35.46	431.47
^{48}Cd	682.70±37.71	650.94	135.67±8.80	129.86	818.37±69.94	780.81
^{68}Er	2434.32±140.60	2303.17	476.18±31.96	570.26	2910.50±277.61	2873.43

K shell fluorescence yields and level widths are given in Table 2 with the theoretical results. The unit of all level widths is eV. As seen from Table 2, the agreement of K shell fluorescence yield (ω_K) between the present results and theoretical predictions of Krause [1] are within the range <4.8%. The agreement of level widths between the measured values and theoretical predictions ([9], [10]) are < 7.6%, ≤ 8.1% for K shell, respectively. As seen from Table 2, the measured $\omega_K$and Γ_K values increase with increasing atomic number.The error in ω_K and Γ_K values is less than < 9.4% for all targets and it mainly arises from the experimental errors in K shell X-ray fluorescence cross sections. The present results are generally in good agreement with other theoretical predictions and theoretical values present in the literature.

Table 2. Experimental and theoretical K shell X-ray fluorescence yields (ω_K) and level widths (Γ_K) values.

Element	ω_K		Γ_K **(eV)**	
	Experimental	Theoretical	Experimental	Theoretical
^{40}Zr	0.749±0.060	0.730 [1]	3.59±0.28	3.84[9] 3.83[10]
^{42}Mo	0.791±0.063	0.765[1]	4.21±0.33	4.52[9] 4.52[10]
^{48}Cd	0.883±0.075	0.843[1]	6.73±0.57	7.28[9] 7.32[10]
^{68}Er	0.959±0.086	0.947[1]	27.39±2.61	28.4[9] 28.5[10]

For the measured and calculated radiative vacancy transfer probabilities, eq. (9) were used to include partial radiative transition rates (Scofield, [23]) and level widths from Table 2.The experimental values of K to L_2, L_3, M_2, M_3 radiative vacancy transfer probability are compared with the calculated values from Krause and Oliver [9] and Campbell and Papp [10] in Table 3.The agreement of radiative vacancy transfer probabilities between the measured values and theoretical predictions from [9], [10] are < 8.5%, ≤ 9% for K shell, respectively.

Table 3. Experimental and theoretical K shell to L_2, L_3, M_2, M_3 sub-shells radiative vacancy transfer probability values.

Element	η_{KL2} (R)		η_{KL3} (R)	
	Experimental	Theoretical	Experimental	Theoretical
^{40}Zr	0.219±0.017	0.205[a]	0.419±0.033	0.392[a]
		0.205[b]		0.393[b]
^{42}Mo	0.230±0.018	0.215[a]	0.439±0.035	0.409[a]
		0.215[b]		0.409[b]
^{48}Cd	0.256±0.022	0.236[a]	0.480±0.041	0.444[a]
		0.235[b]		0.441[b]
^{68}Er	0.277±0.026	0.267[a]	0.492±0.047	0.474[a]
		0.266[b]		0.473[b]
Element	η_{KM2} (R)		η_{KM3} (R)	
	Experimental	Theoretical	Experimental	Theoretical
^{40}Zr	0.0330±0.0026	0.0309[a]	0.0644±00051	0.0602[a]
		0.0310[b]		0.0603[b]
^{42}Mo	0.0358±0.0028	0.0333[a]	0.0696±0.0055	0.0648[a]
		0.0333[b]		0.0648[b]
^{48}Cd	0.0424±0.0036	0.0392[a]	0.0824±0.0070	0.0761[a]
		0.0390[b]		0.0757[b]
^{68}Er	0.0512±0.0049	0.0494[a]	0.0990±0.0094	0.0955[a]
		0.0492[b]		0.0952[b]

[a]Theoretical calculation of radiative vacancy transfer probability from eq. (9) obtained theoretical level widths from Krause and Oliver [9]

[b]Theoretical calculation of radiative vacancy transfer probability from eq. (9) obtained theoretical level widths from Campbell and Papp [10]

The present results are generally in the good agreement with theoretical prediction and other experimental results can be obtained from literature. In order to have a further check for the accuracy and reliability of measured values, more experimental data are needed. The present results confirm the XRF method is simple, direct and fast and reliable way acquiring information on elemental analysis.

References

[1] MOKrause and JPhys1979Chem. *Ref. Data***8** pp 307
[2] RYilmaz, HTunç and AÖzkartal2015*Radiat. Phys. Chem.***112** pp 83
[3] ÖŞimşek, ODoğan, ÜTurgut and MErtuğrul2000*Radiat. Phys. Chem.***58** pp 207
[4] RDurak and YÖzdemir2001*Radiat. Phys. Chem.***61**pp 19
[5] IHan, MŞahin, LDemir and YŞahin2007*Appl. Radiat. Isot.***65** pp 669
[6] FAkman, RDurak, MRKaçal and MFTurhan2015*Can. J. Phys. 93* pp 1057
[7] SBGudennavar, NMBadiger, SRThontadarya and BHanumaiah2003*Radiat. Phys. Chem.***68** pp 721
[8] RDurak1998Physica *Scripta.***58** pp 111
[9] MOKrause and JHOliver1979 *J. Phys. Chem. Ref. Data* **83**pp 329
[10] JL Campbell, T Papp2001 *At. Data Nucl. Data Tables* **77** pp 1
[11] ECengiz, E Tıraşoğlu, G Apaydın, V Aylıkcı, N Küp Aylıkcı and CAksoy2011*Radiat. Phys. Chem.***80** pp 328
[12] M Ertugrul2002 *J. Anal. At. Spectrom.***17** pp 400

[13] M Ertuğrul2002 *Anal. Chem. Acta* **454** pp 327
[14] F Akman2016Appl. *Radiat. Isot.***115** pp 295
[15] AS Bennal, KMNiranjan and NMBadiger2010 *J. Quant. Spectrosc. Radiat. Transfer***11** pp 1363
[16] JReyes-Herrera, JMiranda2008*Nucl. Instrum. Methods Phys. Res., Sect. B***266** pp 5075
[17] EV Bonzi2006 *Nucl. Instrum. Methods Phys. Res., Sect. B***245** pp 363
[18] FTuzluca, ÖSöğüt, EBüyükkasap, BGDurdu and AKüçükönder2008 *Radiat. Phys. Chem.***77** pp 996
[19] MF Turhan, R Durak and F Akman2014 *Appl. Radiat. Isot.* ***89*** pp 151
[20] LGerward, NGuilbert, KBJensen and HLevring2001 *Radiat. Phys. Chem.***60** pp 23
[21] MJ Berger, JH Hubbell, SM Seltzer, JS Coursey and DSZucker1998*NYST Stand. Ref. Database* 8 (XGAM)
[22] JH Scofield1973 T*heoretical Report no. UCRL. Lawrance Livermore Laboratory, Liver-more, CA p.* 51326
[23] JH Scofield1974At. *Data Nucl. Data Tables* **14** pp 121

IOP Conf. Series: Materials Science and Engineering **282** (2017) 012016 doi:10.1088/1757-899X/282/1/012016

Enhanced definition and required examples of common datum imposed by ISO standard

Yiqing Yan[1,*] and Martin Bohn[2]

[1]Tolerancing and Dimensioning Body-in-White, Mercedes-Benz Cars, Daimler AG, Sindelfingen, Germany
[2]Tolerancing and Dimensioning Body-in-White, Mercedes-Benz Cars, Daimler AG, Sindelfingen, Germany

[*] Corresponding author : yiqing.yan@daimler.com

Abstract. According to the ISO Geometrical Product Specifications (GPS), the establishment and definition of common datum for geometrical components are not fully defined. There are two main limitations of this standard. Firstly: the explications of ISO examples of common datums are not matched with their corresponding definitions, and secondly: a full definition of common datum is missing. This paper suggests a new approach for an enhanced definition and concrete examples of common datum and proposes a holistic methodology for establishment of common datum for each geometrical component. This research is based on the analysis of physical behaviour of geometrical components, orientation constraints and invariance classes of datums. This approach fills the definition gaps of common datum based on ISO GPS, thereby eliminating those deficits. As a result, an improved methodology for a fully functional defined definition of common datum was formulated.

1. Introduction

ISO 5459:2013 [1] is the international Geometrical Product Specifications (GPS) [2] standard for definition and establishment of common datum of geometrical components in design and manufacturing. In ISO 5459:2013 the examples are not matched with the corresponding definitions of common datums. Moreover, there are only two kinds of examples for common datum establishments with the same invariance class of geometrical components: one is with only parallel planes and another is with only parallel axes. It causes an incomplete definition of common datum establishment for geometrical components containing nonparallel planes and axes. This paper forces on the development of an unambiguous definition and an improved holistic methodology for function-oriented establishment of common datum for every geometrical component. This research is based on the analysis of physical behaviour of geometrical components, orientation constraints and invariance classes of datums.

The previous works [3-12] concentred more on the mathematical filtration process and practical application of single datums and datum systems, rather than the development of explicit and complete definition of common datum and function-oriented common datum establishment process.

This paper proposes an enhanced definition and required examples of common datum and gives a new holistic approach for the establishment process of common datum. This paper is constructed as follows: section 2 describes the state of the art including basic theory and terminologies of common datum for geometrical parts from the current ISO standards and examples from ISO with their corresponding deficits. A new enhanced definition with corresponding examples and an improved holistic methodology are given in section 3. The conclusion and outlook is written in section 4.

2. State of the art and deficits

2.1. Terminology

In order to understand the research and purpose of this paper, the following important terminologies are explicated with the definitions and the illustrations:

- *Situation feature:* point, straight line, plane or helix from which the location and orientation of features, or both, can be defined [1].
- *Datum feature:* real integral feature used for establishing a datum [1].
- *Datum feature indicator:* single features to be used for establishing datum features shall be indicated [1]. The symbol is shown in Figure 1.

Figure 1. Datum feature indicator: a) box linked to a filled, b) open datum triangle by a leader line [1].

- *Datum:* is a theoretically exact reference; It is defined by a plane, a straight line or a point, or a combination thereof [1].
- *Common datum*: datum established from two or more datum features considered simultaneously. To define a common datum, it is necessary to consider the collection surface created by the considered datum features [1].
- *Collection surface*: two or more surfaces considered simultaneously as a single surface. When two intersecting planes are considered simultaneously as a single surface, that surface is a collection surface [1].
- *Invariance class:* group of ideal features for which the nominal surface is invariant for the same degree of freedom. The number of invariance degrees is equal to the number of unconstrained degrees of freedom for a given geometrical feature [1]. All geometrical component has 6 degree of freedom (DOF) which contains 3 translations in X, Y and Z direction and 3 rotations around each axis [13-15]. For example: the invariance class of a sphere is spherical. Its datum is the middle point and constrains 3 translations; the invariance class of a plane is planar. Its datum is a plane and constrains 1 translation and 2 rotations; the invariance class of a cylinder is cylindrical. Its datum is the straight line (axis) and constrains 2 translations and 2 rotations [1].

2.2. Examples from ISO: definitions and establishments of common datums

2.2.1. Example I: common datum with two parallel planes.

Figure 2 illustrates the establishing process of common datum by two parallel planes imposed by ISO example. Single parallel datum plane A and B are in part a) and their common datum plane A-B is the middle plane of A and B which shows in c).

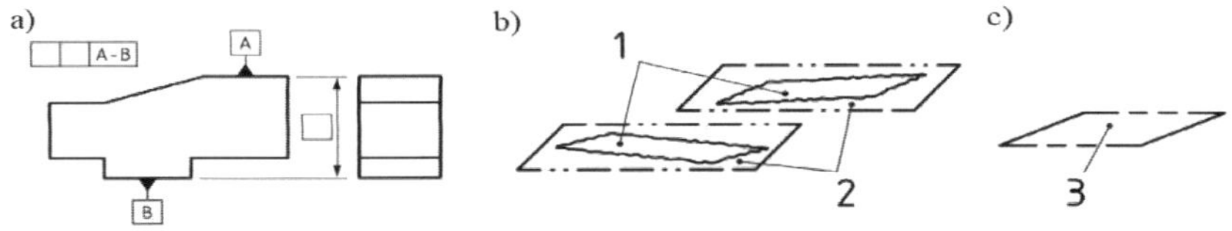

Figure 2. a) two parallel datum planes A and B, b) key 1: datum features, key 2: associated features [16]: theoretical exact single datum planes A and B, c) key 3: common datum- situation feature of the collection surface is the middle plane A-B [1].

2.2.2. Example II: common datum with two parallel axes.

Figure 3 illustrates another example from ISO for establishment of common datum from two parallel datum axes. The single datums of two cylinders are axes A and B. Based on the ISO explanation of this example, their common datum A-B is a plane and a straight line (see Figure 3 c). The plane contains two axes and the straight line is the median straight line of these two axes [1].

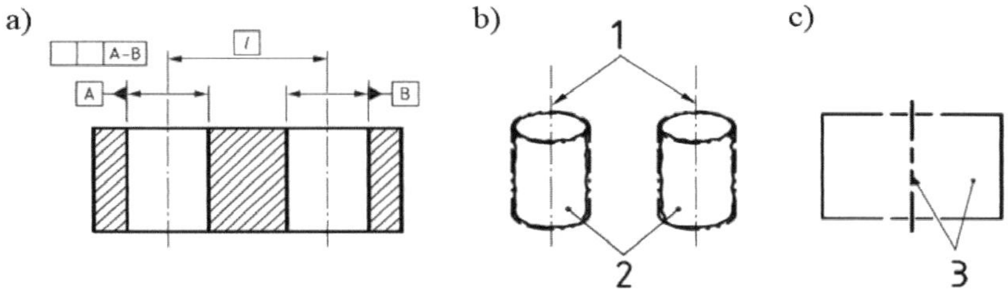

Figure 3. ISO example of common datum with two parallel axes. a) two parallel datum axes A and B, b) key 1: single datum axes A and B, key 2: associated features: theoretical exact cylinders, c) key 3: common datum-situation feature of the collection surface: a plane and a straight line [1].

2.3. Deficits of ISO examples

Based on the ISO definitions of common datum and collection surface in section 2.1 and the explanations of ISO examples in section 2.2, there are two main deficits to be recognised.

Deficit 1: according to the definition of common datum, two datum features in Example II should be considered simultaneously and under the collection (bisection) process. Thereby the common datum should be the median straight line of both datum axes from these two datum features. This is contradictory to the ISO explanation in Figure 3c) which is a plane and a line. Moreover, the ISO definition defines the necessity of consideration of collection surface during common datum establishment. Example II defines the collection surface of these two datum axes which is a plane and a straight line. It is also a contradiction to the original definition of collection surface which is only a surface without straight line.

Deficit 1: The explanation of Example II does not fit the ISO definition of common datum and collection surface. Straight line and plane are not collection surface and the median straight line is not the common datum A-B.

Another deficit is, there are only parallel datum features with the same invariance class (planes and axes) that are taken into account in the ISO examples. A datum is defined by a plane, a straight line and a point. A systematic methodology for definitions and establishment of common datums with concrete examples for the non-parallel datum features including all the situation features of datum are missing.

Deficit 2: The definitions and examples of establishments of common datums imposed by ISO are not complete. A holistic methodology of common datum establishment of planes, straight lines and points is missing.

3. New approach

This section proposes a new approach by analysis of physical behaviour of geometrical components, orientation constraints / DOF between datums and invariance class of datum features. The following section 3.1 enhances the definitions of terminologies of common datum and collection surface. Section 3.2 gives a holistic methodology to improve the completeness of establishment of common datum by using concrete examples. Section 3.2 subclassifies the holistic methodology into three catalogues

IOP Conf. Series: Materials Science and Engineering **282** (2017) 012016 doi:10.1088/1757-899X/282/1/012016

based on the situation features of datum which are a plane, a straight line (in this paper, it is an axis of a cylinder) and a point (in this paper, it is a middle point of a sphere).

3.1. Expansion of definition of common datum

In order to eliminate the Deficit 1 where the definitions of common datum and collection surface do not match the explanations of ISO examples, a function-oriented and enhanced definitions thereof should be added. The function of a common datum of a geometrical component should constrain the same DOF which are constrained through the single datums. The invariance class of a common datum can be the same invariance class of each single datums, but it is not a must (another invariance class is permitted). Important is that the invariance class of a common datum constrain the same DOF which the single datums constrain. It is not necessary to consider the collection surface of the single datums, because not all the common datum of geometrical component is a surface. The thought model of a collection surface can be only applied by the establishment of a common datum by a geometrical component with two parallel single datum planes. Based on the above reasons, the following definition of a common datum is enhanced:

Improvement 1: Enhancement of definition of common datum. *Common datum*: To define a common datum, it is not always necessary to consider the collection surface created by considered datum features. The result of a common datum should not constantly created through the simple collection process of each single datum features. The established common datum should constrain the same degree of freedom as the single datum features collaborative constrain.

3.2. Holistic methodology of definition of common datum by planes, straight lines and points

This paper proposes a completeness of definition of common datum with the geometrical components which are frequently applied in practice. Based on the definition of ISO, datum is distinguished by a plane, a straight line and a point. In practice, datum straight line is mostly considered as an axis of a cylinder and datum point is considered as a middle point of a sphere. Therefore, subsections 3.2.1-3.2.3 give three corresponding holistic methodologies of definitions and examples of common datums separately by planes, axes and middle points. Nevertheless, the corresponding examples do not have just parallel datum features in one geometrical component.

3.2.1. A holistic methodology of definitions and examples for "plane-plane" common datum.

For a geometrical component with two planes as datum features, there are three situations of relative orientation of datum feature planes: parallel, perpendicular and with a defined angle. Figure 4 illustrates the holistic methodology of definitions and examples of the "plane-plane" common datum in each cases. When two datum feature planes are parallel, their common datum plane A-B is the bisector plane of A and B which is defined by ISO. Based on ISO, the notation of A-B and B-A have the identical meaning. When these two planes are perpendicular, the common datum A-B is the combination of datum plane A and a straight line in A. This straight line is the intersection line of plane A and plane B. In this case, A-B has a different meaning to B-A. This methodology defines, when A is notated in the first place of A-B, then A is the datum plane in the common datum. If B-A is written, then B is defined as the datum plane in the common datum. These both single datum planes A and B constrain totally 5 DOF and the new functionally established common datum A-B constrains 5 DOF as well. If the establishment of common datum based on the original definition of ISO in section 2.1, the common datum A-B should be the bisector plane of A and B which is a 45° inclined plane.. This is false because the constrained DOF of a plane is only three. Analog to the perpendicular case, the third case with a defined angle have the same common datum A-B as in the second case. The common datum A-B is a combination of datum plane A (because A is notated in the first place of A-B) and the intersection line of plane A and B. In this case, the common datum A-B and B-A have different meaning in analogy to case II.

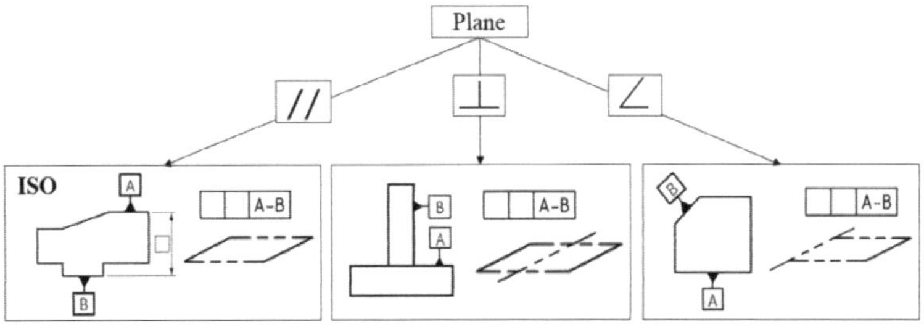

Figure 4. Holistic methodology of "plane-plane" common datum A-B by geometrical component with two datum feature planes A and B. Case I (left): parallel planes; Case II (middle): perpendicular planes; Case III (right): planes with defined angles.

3.2.2. A holistic methodology of definitions and examples for "axis-axis" common datum.
Figure 5 illustrates the holistic methodology of establishment of "axis-axis" common datum by geometrical components with three cases of relative orientation of two single datum axes A and B. Case I is a part with two parallel axes A and B, the common datum A-B is a combination of a plane which contains both axes and a straight line. The straight line is a virtual middle line of both axes. The notations of A-B and B-A have the identical meaning, because of the parallelism of A and B. This is defined by ISO. Case II shows a part with two perpendicular cylinders with two datum axes A and B. They constrain all the DOF of this part where the part is now fixed in space. The common datum A-B is a combination of a plane, a straight line and a point which block all the DOF of this part. The plane contains axes A and B, the straight line is the axis A and the point is the intersection point of axes A and B. In this case, the methodology defines that the definition of A-B is different with B-A, because of the non-parallelism of axes A and B. When the common datum is notated as B-A, so the straight line in the common datum is the axis B instead of axis A. Analog to case II, case III with two cylinders by a defined angle has also the same common datum A-B with a combination of a plane, a straight line (axis A) and a point. The establishing process of the plane, straight line and point is the same as in case II. The meaning of common datum A-B and B-A is also different in analogy with case II.

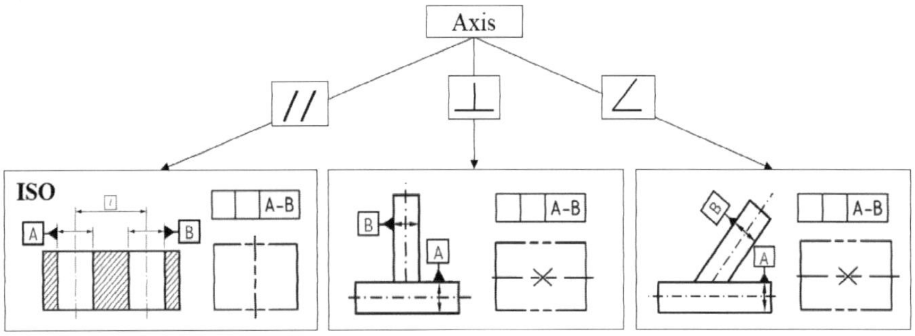

Figure 5. Holistic methodology of "axis-axis" common datum A-B by geometrical component with two cylindrical datum features A and B. Case I (left): parallel axes; Case II (middle): perpendicular axes; Case III (right): axes with defined angles.

3.2.3. A holistic methodology of definition and example for "point-point" common datum.
The example and definition of establishment of common datum by a geometrical component with spheres is completely missing in the current ISO. Figure 6 gives a definition and example of common datum by two spherical datum features A and B. Mathematically there is no parallelism and perpendicularity between two points. Therefore, theoretically the description of relation between two points in space is named: relative orientation. Figure 6 shows a component with two single datum middle points A and B, their common datum A-B is a combination of a straight line (which is the

connection of middle points A and B) and the virtual middle point (centring point) of A and B. The common datum A-B allows this part only one DOF which is the rotation around the straight line. This is the same as when the single datum points A and B have collaborative constrain. The meaning of common datum A-B and B-A is the same in this case, because of the relative orientation statues which include the interpretation of parallelism.

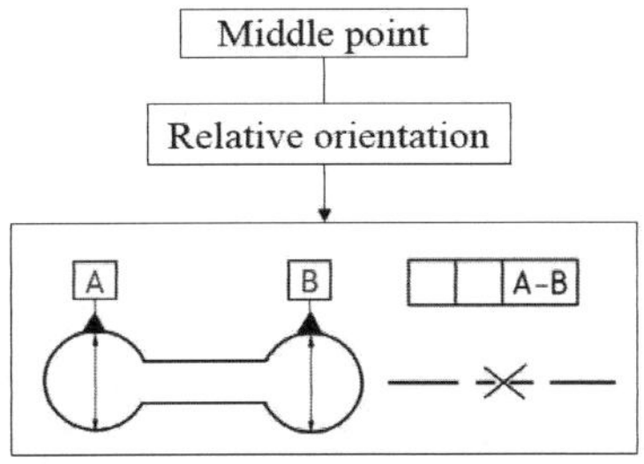

Figure 6. Holistic methodology of "point-point" common datum A-B by geometrical component with two spherical datum features A and B. The common datum is a combination of a straight line and a point.

Section 3.2 describes a function-oriented, clarified and fully defined holistic methodology, in which the geometrical components with two relative oriented planes, straight lines and points are explicated with corresponding practical examples. The definition of common datum notation A-B and B-A are also given in this section. This section eliminates the Deficit 2.

Improvement 2: Complement of definition of common datum. A holistic methodology of definitions and examples of common datum is developed.

4. Conclusion and outlook

This paper presented a novel approach for the enhanced definition and required examples of common datum imposed by ISO. A holistic methodology of establishment of common datum by geometrical components with different oriented planes, straight lines and points were developed with concrete examples. This new approach eliminated the deficits and limitations imposed by ISO standard. Moreover, a new thought model of establishment of common datum by considering the DOF rather than collection surface was formulated as well. Future work involves defining the definition of common datum by mixed combination of planes, straight lines (not only axes) and points (not only middle points). Nevertheless, further work will also contain a study on more than 2 single datums by establishing a common datum. A hypothesis to a complex common datum like A-B-C with different invariance classes as datum features in one geometrical component is: the common datum should constrain the same DOF as the single datums collaborative constrain.

References

[1] 2013 International Standard Organisation, DIN EN ISO 5459, Geometrical Product Specification (GPS) – *Geometrical tolerancing- Datums and datum systems*

[2] Z Huminenny 2009 State of art in standardization in GPS area. *CIRP Journal of Manufacturing Science and Technology* Vol **2** Issue 1 pp 1-7

[3] B Anselmetti and H Louati 2005 Generation of manufacturing tolerancing with ISO standards, *International Journal of Machine Tools and Manufacture* Vol **45** Issue 10 pp 1124-31

[4] V Bhat and E C De Meter 2000 An analysis of the effect of datum-establishment methods on the geometric errors of machined features, *International Journals of Machine Tools & Manufacture* Vol **40** Issue 13 pp 1951-75

[5] P Green 2005 *The geometrical tolerancing desk reference. Creating and interpreting ISO standard technical drawings, (1st ed, Oxford, Elsevier)*

[6] G Henzold 2006 *Geometrical dimensioning and tolerancing for design manufacturing and inspection* (2nd ed, Oxford Butterworth-Heinemann)
[7] C Shakarji and V Srinivasan 2016 *Theory and algorithm for planar datum establishment using constrained total least-squares 14th CIRP CAT* Vol **43** pp 232-37
[8] X Zhang and U Roy 1993 Criteria for establishing datums in manufactured parts *Journal of Manufacturing Systems* Vol **12** No 1 pp 36-50
[9] C Rastogi, R Frutiger, P Yoon and H Lin 2001 *Math based datum selection and optimization for body panels and subassemblies SAE technical paper series* pp 01-3065
[10] W Jordan and W 2014 *Schütte, Form- und Lagetoleranzen: Handbuch für Studium und Praxis, (Munich: Hanser)*
[11] Y Wu and Q Gu 2016 *The Composition Principle of the Datum Reference Frame Procedia CIRP* Vol **43** pp 226-31
[12] S Cheraghi, G Jiang and J Ahmad 2003 An efficient approach for the identification of candidate datums set for a nominally flat primary datum feature *CIRP International Journal of Machine Tools and Manufacture* Vol **43** Issue 4 329-36
[13] 2009 The American Society of Mechanical Engineers, ASME Y14.5 *Dimensioning and Tolerancing- Engineering Drawing and Related Documentation Practices*
[14] M Bohn and K Hetsch 2016 *Funktionsorientiertes Toleranzdesign (Munich, Hanser)*
[15] M Bohn and K Hetsch 2013 *Toleranzmanagement im Automobilbau (Munich, Hanser)*
[16] 1999 *International Standard Organisation EN ISO 14660-1 Geometrical Product Specification (GPS) – Geometrical features- Part 1: General terms and definitions*

Automated Guided Vehicle For Phsically Handicapped People – A Cost Effective Approach

Dr. G. Arun Kumar[1] and Mr. A. Sivasubramaniam[2]

[1]Professor and Head, Department of Mechanical Engineering, Sathyabama University, Chennai, Tamilnadu, India.
[2]Assistant Professor, Department of Mechanical Engineering, VSB Engineering College,Karur, Tamilnadu, India.

*E-mail: garun55@gmail.com

Abstract. Automated Guided vehicle (AGV) is like a robot that can deliver the materials from the supply area to the technician automatically. This is faster and more efficient. The robot can be accessed wirelessly. A technician can directly control the robot to deliver the components rather than control it via a human operator (over phone, computer etc. who has to program the robot or ask a delivery person to make the delivery). The vehicle is automatically guided through its ways. To avoid collisions a proximity sensor is attached to the system. The sensor senses the signals of the obstacles and can stop the vehicle in the presence of obstacles. Thus vehicle can avoid accidents that can be very useful to the present industrial trend and material handling and equipment handling will be automated and easy time saving methodology.

1. Introduction

An automated guided vehicle or automatic guided vehicle (AGV) is a mobile robot that follows markers or wires in the floor, or uses vision or lasers [1]. The AGV can tow objects behind them in trailers to which they can autonomously attach. Automated guided vehicles increase efficiency and reduce costs by helping to automate a manufacturing facility or warehouse [2]. AGVs are employed in nearly every industry, including, pulp, paper, metals, newspaper, and general manufacturing. Transporting materials such as food, linen or medicine in hospitals is also done. Automated Guided Vehicle or AGV is one of material handling equipment that has been widely used in most manufacturing industry today as it provides more flexibility to the systems, the design of this AGV has to be done properly in order the maximize the usability of this AGV [3-4].

The mechanical part or system of this AGV will be incorporated with the guidance system done. The basic concept of the AGV incorporates battery-powered and driverless vehicles with programming capabilities for path selection and positioning [5].

The software used for analysis in this project is FEMPRO V22. It is used to find the displacement for this AGV when the load is applied to the base structure and other components [6]. The choice of the material for the structure is also important as it will compromise the weight to power ratio of this vehicle.

The steering system also will be one of the important of AGV as it will contribute to the radius of turn angle and also the design of the AGV focusing on the mechanical design concept of the AGV which combines knowledge on mechanical parts such as the electric motor, gears, wheels, structure of the AGV and others mechanical parts that are essential [7].

The AGV operated in a model warehouse, built to scale. Its primary task was to relocate pallets within the warehouse. An external input generated by an infrared remote Control notified the AGV whether a pallet was entering or exiting the warehouse [8]. When the four pair line follower module detected an intersection, the AGV determined whether to turn or go straight by using an algorithm that incorporated the vehicles current location and direction. A radio frequency remote was used to control the AGV signalling another autonomous vehicle, the ASRS, the new location of the pallet. The second operation allowed for a pallet to be shipped out of the warehouse [9-10].

The AGV waited for the ASRS to confirm that it dropped off a pallet at one of the transition docks before it picked it up and moved the pallet to the outgoing shipping dock [11]. While travelling, the AGV polled two forward facing infrared range finding sensors to determine if an object was in its forward collision path. If any object was detected, the Vehicle would stop and wait for the obstruction to be cleared. While reversing, bump Sensors detected the occurrence of a rear collision [12].

2. Description of Automatic Guided Vehicle

The Automatic Guided Vehicle used for material handling in a industry is shown in Figure 1. The components include chassis and the steering system [1]. Functions of chassis is to act as a frame for attaching other components,Carry the load of other components and the payload,Act as sacrificial component to prevent damage of expensive payload in case of accidents. Electrical components include the motor and the power supply for the motor itself [2]. Battery is being used. Electronic components provide sensing, logical decision and control of the vehicle. It includes programmed microcontroller for the decision logic, the motor driver as both sensing and control of motor [3]. The predetermined path of the AGV is shown in Figure 2.

Figure 1. The AGV.

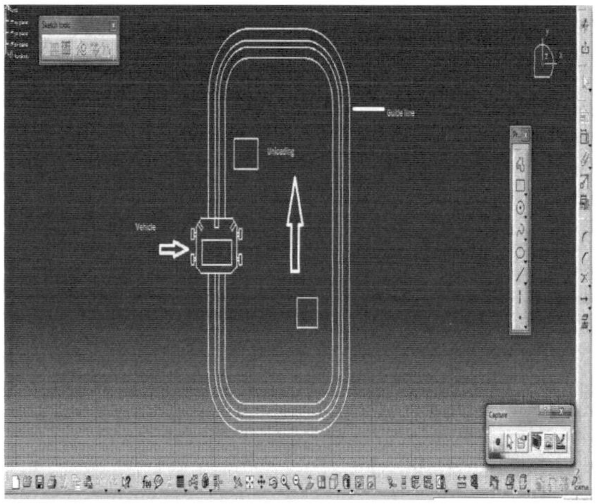

Figure 2. Predetermined Path of the AGV.

3. Electrical and Electronic Components

The transmitter receiver pair is shown in Figure 3 that transmits and receives electronic signal. It is used for transmitting the path data to the AGV microcontroller. The transmitter Module has 2 pins, the receiver has 2 pins. Two set of transmitter receiver is being used to track the path of the AGV. Infrared sensors ate being used as receivers [4] .

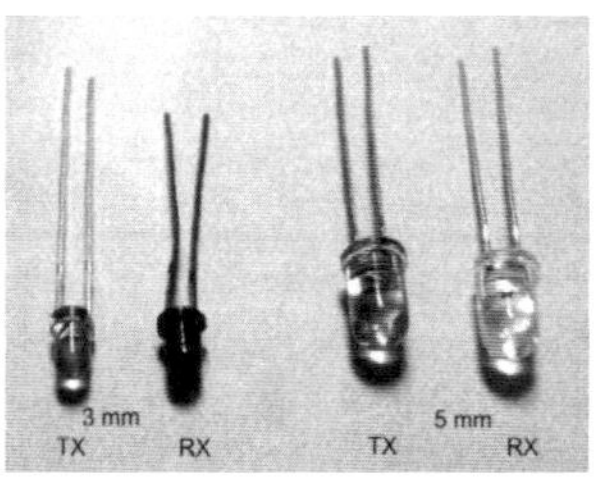

Figure 3. Transmitter and Receiver-Pair.

A microcontroller (µC, uC or MCU) is a small computer on a single integrated circuit containing a processor core, memory, and programmable input/output peripherals. Microcontrollers are designed for embedded applications, in contrast to the microprocessors used in personal computers or other general purpose applications [5]. The Microcontroller used in the AGV is ATMEL 8051 -89S52. The reasons for using ATMEL8051 -89S52 are cheap cost, Easy to program, Availability of UART Communication, High stress values. The Microcontroller is programmed with the required program to accept the data from the transmitter, interpret it, and calculate the path in terms of spatial orientation and five logical values to the Motor driver [6].

The LM158 series consists of two independent, high gains, internally frequency compensated operational amplifiers which were designed specifically to operate from a single power supply over a wide range of voltages. Operation from split power supplies is also possible and the low power supply current drain is independent of the magnitude of the power supply voltage. Application areas include transducer amplifiers, dc gain blocks and all the conventional op amp circuits which now can be more easily implemented in single power supply systems [7]. For example, the LM158 series can be directly operated Off of the standard a5V power supply voltage which is used in digital systems and will easily provide the required interface electronics without requiring the additional g15V power supplies. The UTC ULN2003 is high-voltage, high-current Darlington drivers comprised of seven NPN Darlington pairs. 30RPM Centre Shaft Economy Series DC Motor is high quality low cost DC geared motor is shown in Figure 4. It has steel gears and pinions to ensure longer life and better wear and tear properties[8]. The gears are fixed on hardened steel spindles polished to a mirror finish. The output shaft rotates in a plastic bushing. The whole assembly is covered with a plastic ring. Gearbox is sealed and lubricated with lithium grease and require no maintenance. The motor is screwed to the gear box from inside[9].

Figure 4. DC Geared Motor.

Although motor gives 30 RPM at 12V but motor runs smoothly from 4V to 12V and gives wide range of RPM, and torque. Tables below gives fairly good idea of the motor's performance in terms of RPM and no load current as a function of voltage and stall torque, stall current as a function of voltage[10].

3.1. Atmel 8-Bit Microcontroller

The AT89S52 is a low-power, high-performance CMOS 8-bit microcontroller with 8K bytes of in-system programmable Flash memory is shown in Figure 5. The device is manufactured using Atmel's high-density nonvolatile memory technology and is compatible with the industry-standard 80C51 instruction set and pin out. The on-chip Flash allows the program memory to be reprogrammed in-system or by a conventional nonvolatile memory programmer [11]. By combining a versatile 8-bit CPU with in-system programmable Flash on a monolithic chip, the Atmel AT89S52 is a powerful microcontroller which provides a highly-flexible and cost-effective solution to many embedded control applications. The AT89S52 provides the following standard features: 8K bytes of Flash, 256 bytes of RAM, 32 I/O lines, Watchdog timer, two data pointers, three 16-bit timer/counters,a six-vector two-level interrupt architecture, a full duplex serial port, on-chip oscillator, and clock circuitry. In addition, the AT89S52 is designed with static logic for operation down to zero frequency and supports two software selectable power saving modes[12].

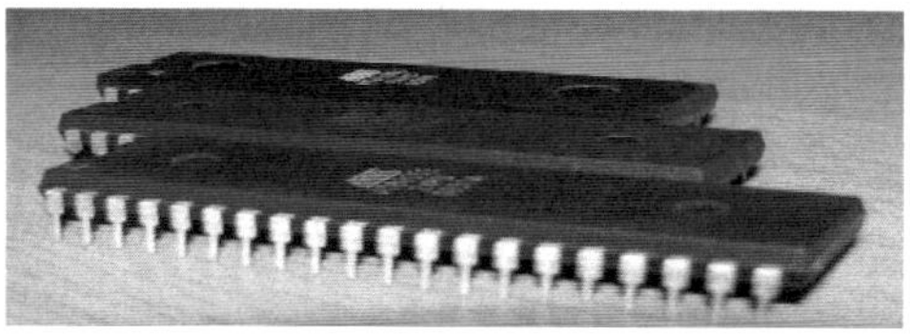

Figure 5. Atmel 8-Bit Microcontroller.

The Idle Mode stops the CPU while allowing the RAM, timer/counters, serial port, and interrupt system to continue functioning. The Power-down mode saves the RAM contents but freezes the oscillator, disabling all other chip functions until the next interrupt or hardware reset. The specifications of the AGV's ATMEL Microcontroller is shown in Figure 6.

Parameter	**Value**
Flash (Kbytes):	8 Kbytes
Max. Operating Frequency	24 MHz
CPU:	8051-12C
Max I/O Pins	32
UART:	1
SRAM (Kbytes):	0.25
Operating Voltage (Vcc):	4.0 to 5.5
Timers:	3
ISP:	SPI

Figure 6. Specifications of ATMEL8051 -89S52 Microcontroller.

3.2. Memory Organization
MCS-51 devices have a separate address space for Program and Data Memory. Up to 64K bytes each of external Program and Data Memory can be addressed [13-14].

3.3. Program Memory
If the EA pin is connected to GND, all program fetches are directed to external memory. On the AT89S52, if EA is connected to VCC, program fetches to addresses 0000H through 1FFFH are directed to internal memory and fetches to addresses 2000H through FFFFH are to external memory.

3.4. Data Memory
The AT89S52 implements 256 bytes of on-chip RAM. The upper 128 bytes occupy a parallel address space to the Special Function Registers. The Data memory is shown in Figure 7. This means that the upper 128 bytes have the same addresses as the SFR space but are physically separate from SFR space. When an instruction accesses an internal location above address 7FH, the address mode used in the instruction specifies whether the CPU accesses the upper 128 bytes of RAM or the SFR space. Instructions which use direct addressing access of the SFR space. For example, the following direct

addressing instruction accesses the SFR at location 0A0H (which is P2).MOV 0A0H, #data Instructions that use indirect addressing access the upper 128 bytes of RAM [15]. For example, the following indirect addressing instruction, where R0 contains 0A0H, accesses the data byte at address 0A0H, rather than P2 (whose address is 0A0H). MOV @R0, #data Note that stack operations are examples of indirect addressing, so the upper 128 bytes of data RAM are available as stack space.

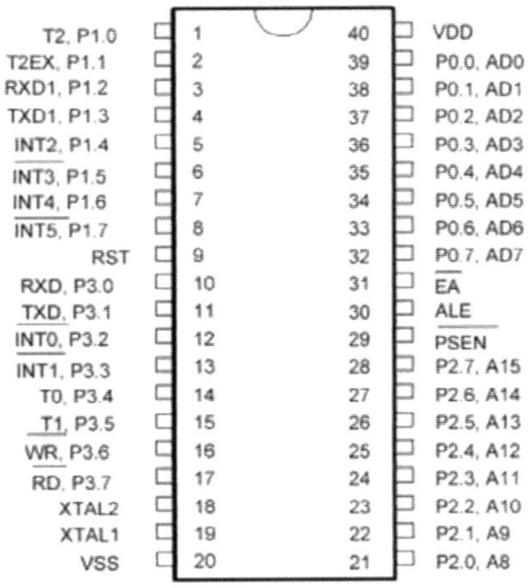

Figure 7. Data Memory.

3.5. Circuit Connection:

The AGV's circuit connection is shown in Figure 8.

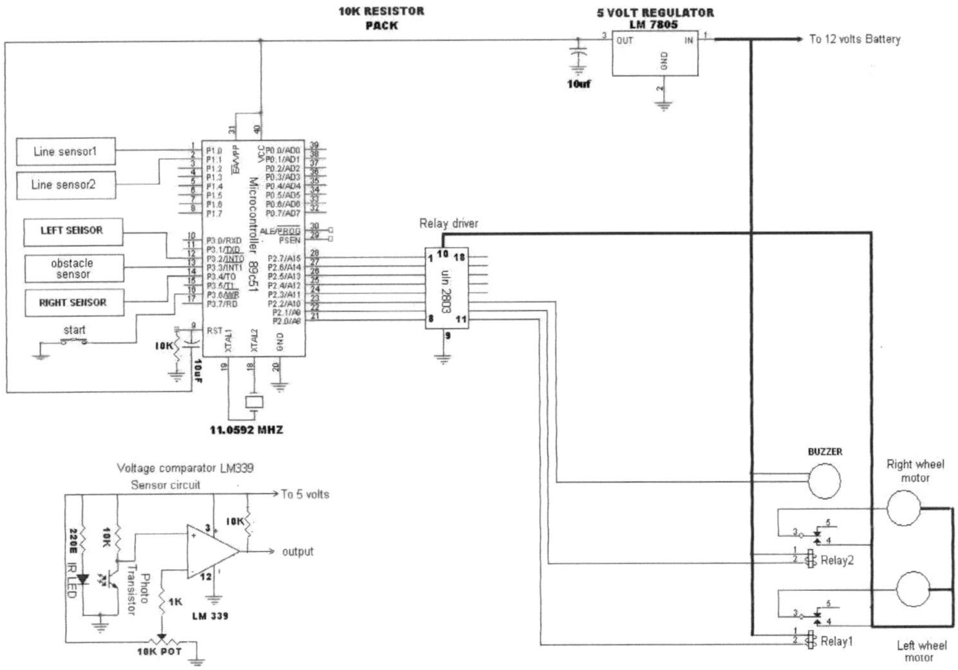

Figure 8. Circuit Connection.

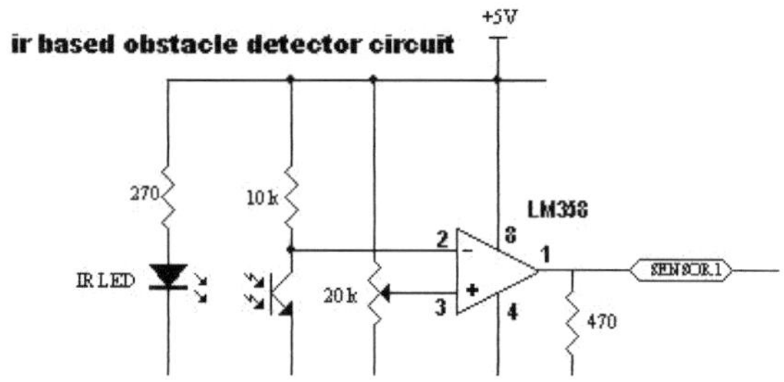

Figure 9. IR based obstacle detector circuit.

The Infrared based obstacle detector circuit diagram is shown in Figure 9. It will sense and detect the obstacle. Then the AGV will deviate from its orgnial path and then it will regain its orginal path after crossing the obstacle.

4. Path Planning

The AGV's movement is based on path planning. The path planning is based on spatial graph theory. It is represented in Figure 10.

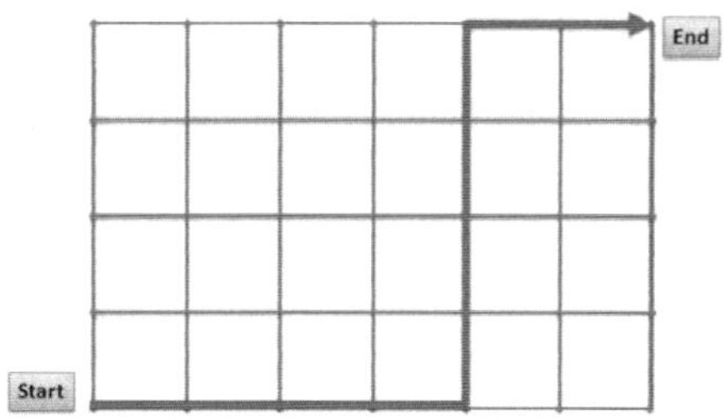

Figure 10. Spatial Graph Theory.

4.1 Spatial Graph Theory

The path is not currently in use by other AGV and the path does not cross the lines being used by other AGV,the path is under maintenance.

5. Cost Analysis of AGV

Estimated cost of the AGV model is Rs 8000/-
The Fabricated model is shown in the Figure11.

5.1 The AGV is a productivity increasing feature in a factory that has the following advantages

- Speed of delivery
- Flexibility of path
- Adaptive to changes in factory layouts
- Reduction in labor cost and Running cost
- Ability to add sensors to detect the payload conditions

5.2 Disadvantages

- Should be recharged periodically

- Will stop delivery when the AGV is forced off the path.
- High Initial cost

Figure 11. Fabricted Model of the Automated Guided Vehicle.

6. Conclusion

The Advantages of the AGV far shadow over the disadvantages and hence it is concluded that in a mass production factory with large area, a AGV will definitely increase productivity, decrease expenditure.

References

[1] Jervis B "Guidance options for AGVs" . Webb Company, 2007.
[2] Mahendra, "Structural and Embedded design", Chrysler Automotive India, 2010
[3] Murty. D.V.S, "Transducers and Instrumentation" Prentice Hall of India Publishing Company Limited, 2001.
[4] Olmi, Roberto (2011). Traffic Management of Automated Guided Vehicles in Flexible Manufacturing Systems. Ferrara (Italy): University of Ferrara.
[5] Paul.B,AGV Drive and Steering Options Transbotics Corp., 2009
[6] Rajneesh Kumar Singh"Automated Trailer Loading AGVs" JBT Corporation. 2009.
[7] R.N.Mall"Pallet Handling AGVs" JBT Corporation. 2009.
[8] Savant "The Basics of Automated Guided Vehicles". AGV Systems.. 2006
[9] Shyam Sundar, "The 8051 Microcontroller Architecture, Programming & Applications", Penram International Publishing (India) Private Limited.
[10] Shyam kumar chowdry,Kuldip kumar sahu.,"Common AGV Applications: Raw Material Handling" JBT Corporation. 2009.
[11] Vivek Kumar,R.N. Mall "Sonar sensor and mounting". University of Birmingham. 2006.
[12] Vivek Kumar,R.N. Mall,Rajneesh Kumar Singh"Towing AGVs ". Hi-Tech Robotic Systemz Ltd. India, 2009.
[13] Vivek Kumar "Common AGV Applications: Roll Handling" JBT Corporation. 2009.
[14] Vivek kumar,R.N.Mall"Battery Charging Systems for Automated Guided Vehicles".AGV Battery Charging Systems. Egemin Automation Inc. 2006.
[15] Vivek kumar"Work in Process Movement with AGVs" JBT Corporation. 2009.

IOP Conf. Series: Materials Science and Engineering **282** (2017) 012018 doi:10.1088/1757-899X/282/1/012018

The effect of fibre loading and graphene on the mechanical properties of goat hair fibre epoxy composite

J Jayaseelan*, K R Vijayakumar, N Ethiraj, T Sivabalan and W Andrew nallayan

Mechanical Engineering Department, Dr M.G.R Educational and Research Institute University, Maduravoyal, Chennai-600095, India

* Corresponding author: see_lan@yahoo.com

Abstract. Composite materials are heterogenous materials containing one or more solid phases. In recent years cost-effective composite making is an ideal task. Hence we have come out with a natural fibre composite, which contains goat hair and epoxy as a binding element, with the combination of Graphene as a main source of enhanced mechanical property. Fabrication of natural composite consists of five layers of goat hair sandwiched in epoxy matrix. These composites made are tested for mechanical properties including Tensile strength, Flexural strength, Inter laminar shear and Impact strength. The mechanical properties of the six composite sets are analyzed and reported.

1. Introduction

As the world of nanotechnology is emerging rapidly, noble awarded material Graphene is highly playing its role in all the field of sciences [1]. Graphene with its extraordinary electrical, optical, mechanical, thermal properties sweeps the industries all over the world [2]. In recent times, composite materials are being used in numerous application fields [2, 3]. Composites are materials consisting two or more chemically distinct constituent on a micro scale having a distinct interface separating them [4, 5]. One or more phase is embedded in a continuous phase to form a composite. The discontinuous phase is usually harder and stronger than the continuous phase and is called the reinforcement, whereas the continuous phase is termed as matrix [6]. Goat hair is a natural fibre which possesses good mechanical properties like tensile strength, resistance to decomposition, reusable resource hence can replace synthetic fibres. Now the combination of goat hair and Graphene enhances all the major properties of a composite material covering its tensile strength, impact tensile strength, flexural strength, impact strength; inter laminar shear strength and hardness [6].

2. Goat hair

Goat hair is a natural fibre extracted from the skin of goat. The goat hair varies between 40 to 120mm in length, diameter between 80 to 105microns, tensile strength 90 to 120MPa, moisture absorption 7%, and elongation at shear 10 to 30%, bio-degradable, eco-friendly, smooth, wear resistant, soft, elastic, low-cost and available in plenty. Even though goat hair is used for many other applications not much work has been tried out using goat hair for making composite. Disposal of goat hair in tannery process is a perennial problem in waste management. The above issues triggered the author to make goat hair composite and to study them by varying the fibre loading.

The method,that was being used in the fabrication of composite material, is followed by several researchers [7, 8]. The goal of this work is to study and develop a low cost natural composite material using goat hair fibre with improved mechanical properties like wear resistance, tensile and impact strength, inter laminar bonding strength and surface flexural rigidity.

3. Experimental set-up

This experimental set up is designed to make six composites, varying the percentage of goat hair and epoxy as shown in Table 1. Finally composite 6 is fabricated by addition 0.1% Graphene into it. Layers of sheets of goat's hair are piled up one over the other with epoxy matrix in between each layer. The mould set is placed in the compression moulding press under pressure, undisturbed for period of 72 hours for proper setting.

Table 1. Material composition for development and processing of composites.

Sample code	Goat hair fibre (%weight)	Epoxy matrix (%weight)	Graphene filler (% weight)
Composite 1	20	80	0
Composite 2	30	70	0
Composite 3	37.5	62.5	0
Composite 4	40	60	0
Composite 5	50	50	0
Composite 6	37.5	62.4	0.1

3.1. Specimen Preparation

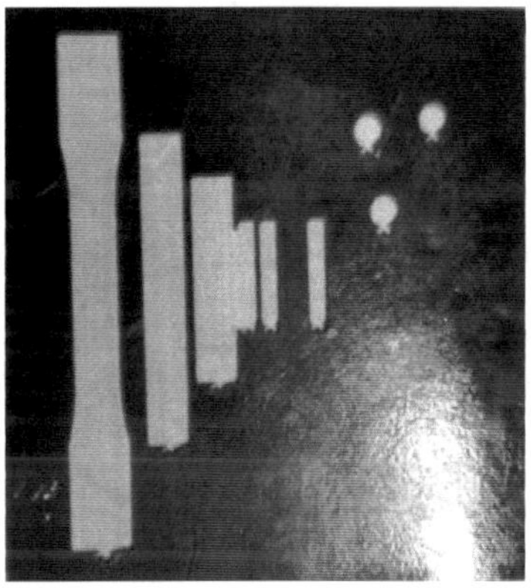

Figure 1. Goat hair epoxy composite

Goat hair epoxy composite fabricated and specimen for tensile testing is shown in figure 1.Composite 1 with 20% goat hair fibre loading is made with five layers of goat hair fibre. Mould plate of 180mm x180mm x8mm thickness dimension is placed are covered with 4mm thick wooden strips and mould release is applied. Epoxy and hardener are mixed in the ratio 10:1 and applied on the mould and then five layers of goat hair are laid over it after applying resin in between each layer. Finally resin is applied on the top of goat hair layer and top mould plate is placed on top and the mould set is kept in compression moulding press, un disturbed for three full days and finally composite 1 is removed from the mould.

Composite 2 with 30% goat hair fibre loading is made by compression molding with five layers of goat hair fibre in epoxy matrix.The composite made is placed under pressure for three days and then removed from mould. Composite 3 with 37.5% goat hair fibre loading is made by compression moulding with five layers of goat hair fibre sandwiched in epoxy matrix.The composite made is placed under pressure for three days and then removed from mould. Composite 4 with 40% goat hair fibre loading is made with five layers of goat hair. Same way, composite 5 with 50% goat hair fibre loading is prepared by 5layers of goat hair. Finally, our target composite 6 is made by addition of 0.1% of Graphene with goat fibres of 5 layers each with 37.5% fibre loading.

4. Results and discussion

4.1. Tensile Test

Tensile strength is a measurement of force required to pull a material to the point where it breaks [9]. A tensile strength of a material is the maximum amount of tensile stress that it can take before breaking. Tensile strength for all the composites was taken. Graphs are drawn for stress vs. strain, load vs. displacement, and stress vs. Displacement [10, 11]. Consecutive reports of the graphs are shown above. The tensile test results are shown in figure 2 and 3.

Therefore, increase in the tensile strength of 36% is abruptly shown in composite 6, due to the addition of Graphene compared with composite 3 showing 33% of tensile strength with 0% of Graphene. Tensile testing, also known as tension testing is a fundamental materials science test in which a sample is subjected to a controlled tension until failure. Composites are tested for tensile strength as per ASTM D638-03 test standard[12].

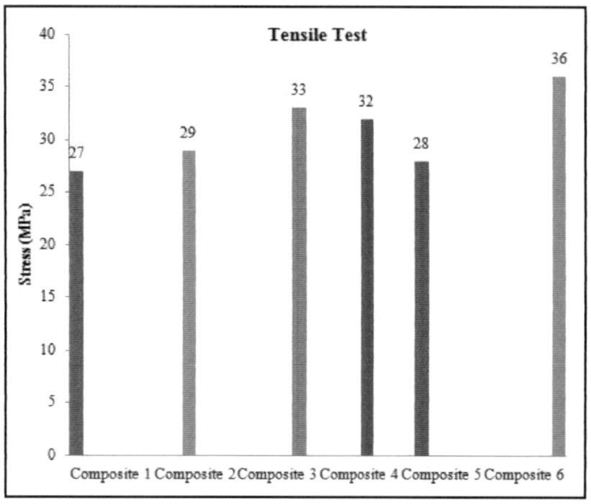

Figure 2. Tensile strength of composites

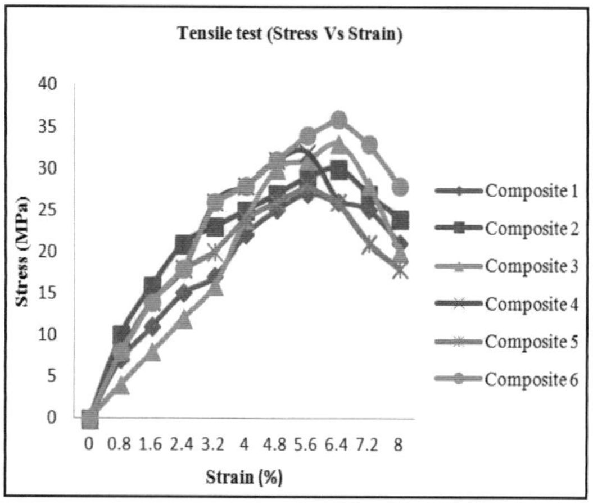

Figure 3. Tensile stress vs % strain

4.2. Flexural Strength

Flexural strength is a measure of an unreinforced concrete beans or slab to resist failure in bending [13,14]. Flexural strength, also known as modulus of rupture, bend strength, or fracture strength, a mechanical parameter for brittle material, is defined as a material's ability to resist deformation under load is conducted as per ASTM D790 test standard [15,16]. The transverse bending test is most frequently employed, in which a specimen having either a circular or rectangular cross-section is bent until fracture or yielding using a three point contact technique. The flexural test results are shown in figure 4 and 5.

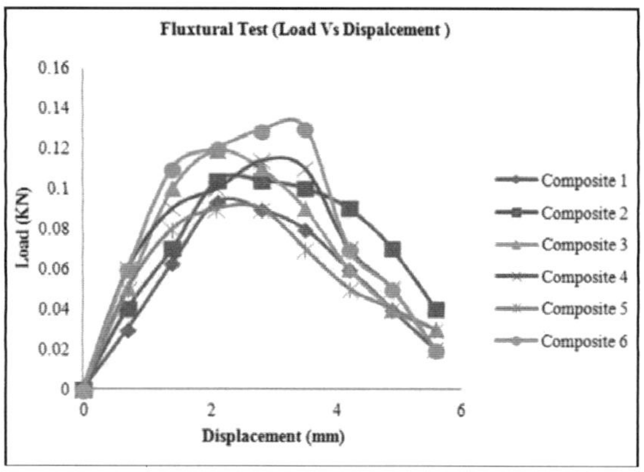

Figure 4. Flexural load vs displacement

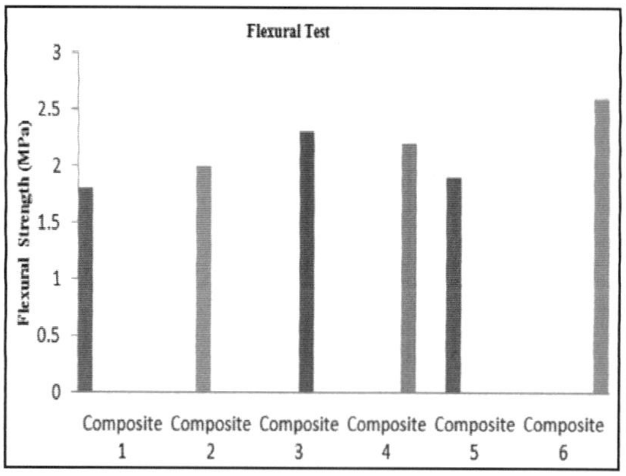

Figure 5. Flexural strength of composites

Composite 1, made by compression moulding, which has got five layers of goat hair with fibre loading of 20% withstands a flexural load of 0.09kN for 5.9mm dispacement ony and it breaks giving a flexural strength of 1.8Mpa. Composite 2, made by compression moulding, which has got five layers goat hair with fibre loading of 30% withstands a flexural load of 0.1N and it started to bend. It is observed that the material break after a displacement of 5.9mm. Thus flexural strength is 2.0Mpa. Composite 3, made by compression moulding, which has got five layers of goat hair and fibre loading of 40%withstands a flexural load of 0.12%KN and then starts to flex. It is observed that the specimen break after a displacement of 5.9mm giving a flexural strength of 2.3Mpa. Thus it increases continuosly . But then a break down of flexural strength takes places due to over fibre load.

Composite4,of fibre loading 50% withstands a flexural load of 0.11KN it starts to flex and breaks down at a displacement of 5.9mm., giving a flexural strenth of 2.2Mpa and composite 5 with a flexural strength of 1.9 Mpa with a load of 0.08KN showing displacement of 5.9mm.Finally by adding graphene ,composite 6 shows a good flexural strength of 2.6 Mpa compared to other composites.

4.3. Impact Strength

The charpy impact test, carried out as per ASTM D256 test standard, also known as the charpy v-notch test, is a standardised high strain rate test which determines the amount of energy absorbed by a material during fracture. This absorbed energy is a measure of composite materials notch toughness and acts as a tool to study temperature-dependent ductile-brittle transition [17,18]. The impact test results are shown in figure 6.

Composite 1, made by compression moulding, which has got five layers of goat hair with fibre loading of 20%, has got Impact strength of 0.54 Joules. Composite 2, made by compression moulding, which has got five layers goat hair with fibre loading of 30%, has got impact strength of 0.64% Joules, which is more than composite1. As the fibre loading increases impact strength also increases. Composite 3, made by compression moulding, which has got five layers of goat hair with fibre loading of 37.5% has got an impact strength of 0.73 Joules. The rise in impact value compared to composite 2 is due to the increase in percentage of fibre loading.Composite4, made up of five layers of goat hair with 40% of fibre loading gives an impact of 0.75%. Composite 5 of fibre loading 50% gives impact energy of 0.79%.Composite 6, with 0.1% of Graphene and fibre loading of 37.5% holds an impact strength of 0.81% which is highest among the six composite.

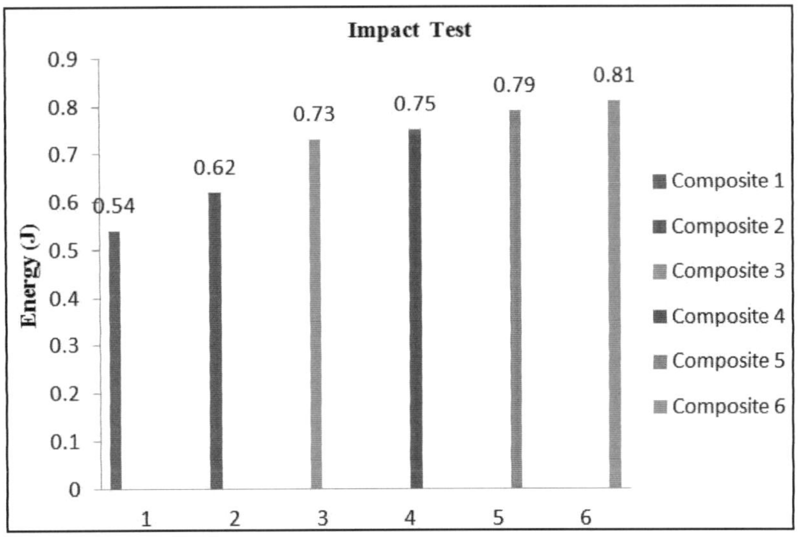

Figure 6. Impact strength of composites

4.4. Inter Laminar Shear Test

Inter laminar shear induces a stress in every composite layer to determine the displacements of the sheets of fibre [19, 20]. The inter laminar shear stiffness relative to the fibre reinforced composite is typically much smaller than the homogenous material. The ILSS test results are shown in Table 2 and figure 7.

Table 2. Inter laminar shear strength of composites

Sample code	ILSS (MPa)
Composite 1	4.50
Composite 2	5.10
Composite 3	6.25
Composite 4	7.30
Composite 5	7.50
Composite 6	8.15

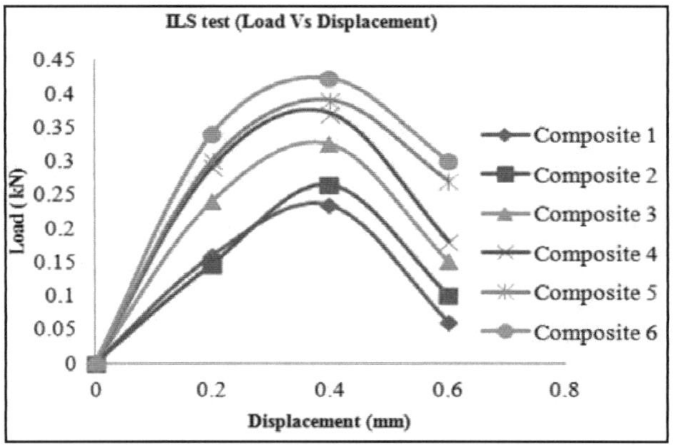

Figure 7. ILSS load Vs displacement

Composite 1 shows shear strength of 4.50 MPa with load 0.2 KN for a displacement of 0.6mm followed by composite 2 showing shear strength of 5.10Mpa with load 0.25KN and displacement of 0.6mm. Composite 3 showing shear strength of 6.25Mpa, at load 0.3 KN. It shows a displacement at a point 0.6 mm. Then composite 4 withstands a load of 0.35 KN for a displacement of 0.6 mm, giving laminar shear of about 7.30Mpa. Composite 5 with a load 0.39 KN at 0.6mm displacement gives a laminar strength of 7.5Mpa. Composite 6 shows a laminar shear at 8.15 Mpa with 0.44 KN load at displacement of 0.6mm.

4.5. Applications
Goat hair fibre epoxy composite can be used to manufacture fibre boats which are very safe even in rough seas and ocean currents. They can be used for noise dampening in generators. They can be used for making helmets. Due to its wear resistance property they can be used for laying roads. They can be used in manufacture of electrical panel boards, doors and windows. They can be used in automobiles for making car bumpers and dash boards. They can be used for manufacturing water tank door lids.

5. Conclusion
Tensile strength of composite increases from 20% fibre loading to 37.5% fibre loading, the maximum value observed being 33MPa. Beyond 37.5% fibre loading the tensile strength drops and finally upon addition of 0.1% graphene in 37.5% fibre loading tensile strength increases from 33 MPa to 36MPa. Fibre loading of 37.5% gave better results in the experiments conducted and addition of graphene increases the tensile strength.

Flexural strength of composite increased from 1.8 MPa to 2.3 MPa upto 37.5% fibre loading, beyond 37.5% fibre loading continuous drop in flexural strength is observed and the flexural strength 2.6 MPa is maximum for Graphene added composite with 37.5% fibre loading.

Impact energy increases from 0.54 Joules to 0.73Joules when fibre loading increases from 20% to 37.5%. Upto 50% fibre loading we could see a small increase in impact strength. Graphene addition increases the strength from 0.73 Joules to 0.81Joules.

Inter laminar shear strength increases from 4.50MPa to 8.15MPa. From 20% fibre loading to 37.5% fibre loading the increase was sharp then beyond 37.5% fibre loading the increase percentage slows down. Adding graphene increases the impact strength from 6.25 MPa to 8.15 Mpa.

Addition of graphene to a smaller extent increases the mechanical properties of the composite and 37.5% fibre loading displays better mechanical properties in the experiment carried out. Further trials could be conducted with hybrid composites to validate the results obtained.

References

[1] J Jayaseelan ,P Palanisamy, K R Vijayakumar and A D M Vinita 2015 *Int. J. Vehicle Struct. and Syst*, **7** pp165-68
[2] J Jayaseelan, P Palanisamy and K R Vijayakumar 2015 *Int.J. Appl. Eng; Res*, **10** pp 689-92
[3] P Wambua, J Ivens and I Verpoest 2003 *Compos. Sci. Tech* **63** pp 1259-64
[4] N Sergio, H Montero, Luiz Augusto, F Terrones , P D Lopes and M JoseRoberto 2005 *Rev. Mater* **10** pp 571-76
[5] D Senthilnathan, A Gnanavel babu, G B Bhaskar and K G S Gopinath 2014 *Int. J. Eng. Tech* **6** pp 75-82
[6] P S Gouda Shivakumar, Raghavendra Kulkarni, S N Kubert and Dayananda Jawali 2013 *Adv. Mater. Lett* **4** pp 261-70
[7] Richard beyak, C F Meyer and G S Kass 1968 *J. Soc. Cosmet. Chem*, **20** pp 615-626
[8] C Popescu and H Hocker 2007 *Chem. Soc. Rev* **37** pp 1282-91
[9] S Negahdari and M Salehi 2012 *Iran. J. Appl. Anim. Sci* **2** pp 27-32
[10] D Jain and A Kothari 2012 *Res. J. recent sci* **1** pp 128-33
[11] K.G.Satyanarayana,, K.Sukumaran , A.G.Kulkarni, S.G.K.Pillai and P.K.Rohatgi , *Compos*, **17**, 329-333 (1986)
[12] M A Dweib, B Hu, A Aidy, H W Shenton and R P Wool 2006 *Compos. Struct* **74** pp 379-88
[13] H Y Cheung, H Meipo, K T Lau, C Francisco and H David 2009 *Compos Part B* **40** pp 655-63
[14] D Chandramohan and K Marimuthu 2011 *Int. J.Recent Res. Appl. Stud* **8** pp 194-206
[15] H L Bros 2002 *J. Mater. sci*, **37**,1683-92
[16] R P Babu, K O'Connor and R Siram 2013 *Prog. Biomater,* **2** pp 1-16
[17] P Abhijit Deshpande, M Bhaskar Rao and C Lakshmana Rao 2000 *J. Appl .Polymer Sci,* **76**, 83-92
[18] K Joseph, S Thomas, C Pavithran and M Brahmakumar 1993 *J. Appl .Polymer Sci* **47** pp 1731-39
[19] A H Zhu, S Murali, W Cai, X Li, J W Suk and J R Potts 2010 *Adv Mater* **22** pp 3906-24
[20] P A Sreekumar, A.Pradeesh, G Unnikrishnan and S Thomas 2007 *Compos. Sci .Tech* **67** pp 453-61

Establishment and analysis of coupled dynamic model for dual-mass silicon micro-gyroscope

Zhanghui Wang[1], Anping Qiu[1,*], Qin Shi[1] and Taoyuan Zhang[2]

[1]School of Mechanical Engineering, Nanjing University of Science and Technology, Nanjing, China
[2]Shanghai Aerospace Control Technology Institute, Shanghai, China

*E-mail:apqiu@mail.njust.edu.cn

Abstract. This paper presents a coupled dynamic model for a dual-mass silicon micro-gyroscope (DMSG). It can quantitatively analyze the influence of left-right stiffness difference on the natural frequencies, modal matrix and modal coupling coefficient of the DMSG. The analytic results are verified by using the finite element method (FEM) simulation. The model shows that with the left-right stiffness difference of 1%, the modal coupling coefficient is 12% in the driving direction and 31% in the sensing direction. It also shows that in order to achieve good separation, the stiffness of base beam should be small enough in both the driving and sensing direction.

1. Introduction

The Dual-mass Silicon Micro-gyroscopes (DSMG) are employing electrostatic actuation and capacitive detection. Limited by the current MEMS fabrication technology, the stiffness of the support beams for the left and right structures of DSMG will be different with the fabrication error [1]. The left-right stiffness difference (stiffness difference) will change the mechanical properties especially the bias and vibration sensitivity of the gyroscope. Therefore, it is important to analyze the impact of stiffness difference on bias and vibration sensitivity of the DMSG.

The fabrication error impacting on the driving direction of the DSMG had been analyzed in some literatures [2-4]. Take into considered of both the drive and sense mode, a non-ideal dynamic model was studied in the literatures [5-6]. However, the stiffness difference of the DSMG in the above mentioned papers was only considered in its driving direction and the modal coupling coefficient was not be proposed.

In this paper, firstly, the coupled dynamic model is proposed for a DMSG. Second, the natural frequencies and modal matrix are calculated. Then the modal coupling coefficient is deduced from the proposed modal matrix. Additionally, the coupled dynamic model can be solved by modal decoupling method. At last, the reliability of the theory is verified by finite element method (FEM) simulation.

2. Structure and operation principle

Fig.1 is the schematic of a dual-mass silicon micro-machined gyroscope. By applying the alternating voltage on the driving comb capacitor, the left and right mass will move in opposite direction along X-axis (driving direction). When the sensor rotates about Z-axis, the resulting Coriolis force causes the left and right mass to move in opposite direction along Y-axis (sensing direction). The relative motion between the movable detection comb and the fixed detection comb forms the differential capacitance for detection. Ideally, the amount of differential detection capacitance is proportional to the input angular rate.

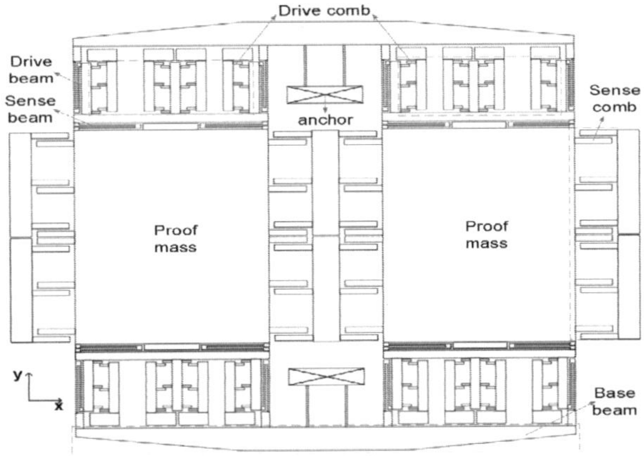

Figure 1. The schematic of a dual-mass silicon micro-machined gyroscope

3. Modeling

Fig. 2 shows the dynamic model in the drive direction and sensing direction. The motion equations are listed as below.

$$m_{d(l,r)}x_{(l,r)} + b_{d(l,r)}\dot{x} + k_{d(l,r)}(x_{(l,r)} - x_b) = F_{d(l,r)} \tag{1}$$

$$m_{bd}\ddot{x}_b + k_{bd}x_b - k_{dl}(x_l - x_b) - k_{dr}(x_r - x_b) = 0 \tag{2}$$

$$m_s\ddot{y}_{(l,r)} + b_{s(l,r)}\dot{y}_{(l,r)} + k_{s(l,r)}(y_{(l,r)} - \theta L_H) = -2\Omega_z\dot{x}_{(l,r)} \tag{3}$$

$$I_b\ddot{\theta} + k_{bs}L_H^2\theta + k_{sl}(y_l + \theta L_H)L_H - k_{sr}(y_r - \theta L_H)L_H = 0 \tag{4}$$

Herein, m_{sl} and m_{sr} are the left and right proof masses, m_{dl} and m_{dr} are the left and right drive comb masses, m_b and I_b are the proof mass of the base beam, $k_{dl}=k_d+\Delta k_d/2$ and $k_{dr}=k_d-\Delta k_d/2$ are the bending stiffness of the left and right drive beams, $k_{sl}=k_s+\Delta k_s/2$ and $k_{sr}=k_s+\Delta k_s/2$ are the bending stiffness of the left and right sense beams. k_{bd} and k_{bs} are the bending stiffness of the base beam in its driving and sensing direction. The parameters of the DMSG designed are list in the Tab. 1.

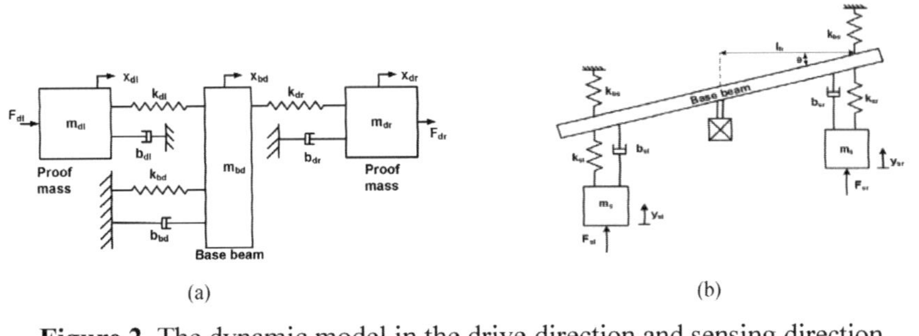

(a) (b)

Figure 2. The dynamic model in the drive direction and sensing direction

Table.1 The parameters of the DMSG designed

Parameter Name	k_b	k_s	Δk_b	Δk_s	k_{bd}	k_{bs}	m_b	m_s
Value	626	674	6.26	6.74	17400	54040	1.08	0.97
Unit	N/m	N/m	N/m	N/m	N/m	N/m	um	um

The base beam mass is much less than the proof mass. Therefore, set m_b and I_b to zero for simplification. Take into account the mass matrix *[m]* and the stiffness matrix *[k]*. Supposing $[z(\omega_n)]=[k-m\omega_n^2)]$, the equations can be written as below.

$$[z(\omega_n)]\{u\} = \{F\}. \tag{5}$$

Then, the natural frequencies can be obtained by setting $|z(\omega_n)|=0$. And the modal matrix [A] defined as the modal matrix is calculated from the eigenvectors of matrix $[z(\omega_n)]$.

$$[A] = \begin{bmatrix} r_1 & r_2 \\ 1 & 1 \end{bmatrix} \tag{6}$$

Here r_1 and r_2 are defined as the mode function. Now, the modal coupling coeffecient which reflects the coupling between the in-phase and anti-phase motion can be express as

$$\alpha_{1,2} = \left|\left|r_{(1,2)}\right| - 1\right| / \left(\left|r_{(1,2)}\right| + 1\right). \tag{7}$$

Finally, the system can be decoupling by applying the modal matrix. As a result the motion equations change to

$$[M_p]\{\ddot{q}\} + [K_p]\{q\} = \{F_p\}. \tag{8}$$

where $[M_p]=[A]^T[M]\,[A]$ and $[K_p]=[A]^T[K]\,[A]$ are the modal mass and modal stiffness. $\{q\}=[A]^T\{u\}$ is the modal freedom. As a consequence, the motion equations can be solved independently.

3.1. Drive mode

The natural frequencies and mode functions in its driving direction are shown as:

$$\omega_{nd}^2 \approx \left(2k_d k_{bd} + 2k_d^2 \pm \left[4k_d^4 + \Delta k_d^2 k_{bd}\left(2k_d + k_{bd}\right)\right]^{1/2}\right)\left(2\left(2k_d + k_{bd}\right)M_d\right)^{-1} \tag{9}$$

$$r_{d(1,2)} = 2k_d^2\left(k_{bd}\Delta k_d \pm \left[4k_d^4 + \left(k_{bd}^2 - 2k_d^2\right)\Delta k_d^2\right]^{1/2}\right)^{-1}. \tag{10}$$

3.2. Sense mode

The natural frequencies and mode functions in its sense direction are shown as:

$$\omega_{sn}^2 = \left(2k_s k_{bs} + 2k_s^2 \pm \left[4k_s^4 + \Delta k_s^2 k_{bs}\left(2k_s + k_{bs}\right)\right]^{1/2}\right)\left(2\left(2k_s + k_{bs}\right)M_s\right)^{-1} \tag{11}$$

$$r_{s(1,2)} = -2k_s^2\left(k_{bs}\Delta k_s + \left[4k_s^4 + \left(k_{bs}^2 - 2k_s^2\right)\Delta k_s^2\right]^{1/2}\right)^{-1}. \tag{12}$$

According to (9)-(12), the natural frequencies and mode functions without the stiffness difference are show as below.

$$\omega_{d\_in-phase}^2 = k_d k_{bd}\left(2k_d + k_{bd}\right)^{-1} M_d^{-1}, \omega_{d\_anti-phase}^2 = k_d / M_d \tag{13}$$

$$\omega_{s\_in-phase}^2 = k_s / M_s, \omega_{d\_anti-phase}^2 = k_s k_{bs}\left(2k_s + k_{bs}\right)^{-1} M_s^{-1} \tag{14}$$

$$\left|r_{d1}\right| = \left|r_{d2}\right| = 1, \left|r_{s1}\right| = \left|r_{s2}\right| = 1 \tag{15}$$

Obviously, the relation in (15) means that the in-phase motion and the anti-phase motion are complete independence. According to (9)-(12), Δk_d and Δk_s are determined by the processing precision. k_d and k_s determine the drive mode and sense mode. As a result, in order to achieve good separation, the stiffness of base beam should be small enough in both the driving and sensing direction. However, these low-rigidity beams present challenges and often introduce other problems.

4. Fem simulation

The Finite element software ANSYS was employed to verify the coupling dynamic model of the DSMG. The schematic of a dual mass silicon micro-machined gyroscope is shown in Fig.3. Frist, the modal analysis of the DSMG structure with stiffness difference is done to verify the natural frequencies. The natural frequency variation due to the stiffness difference of 1% is shown in Tab. 2.

The data in Tab. 2 shows that the natural frequency variations due to the stiffness difference of 1% are 0.5Hz in its driving direction and 1.5Hz in its sensing direction. It also shows that the natural frequency variation can be accurately represented by the coupling dynamic model.

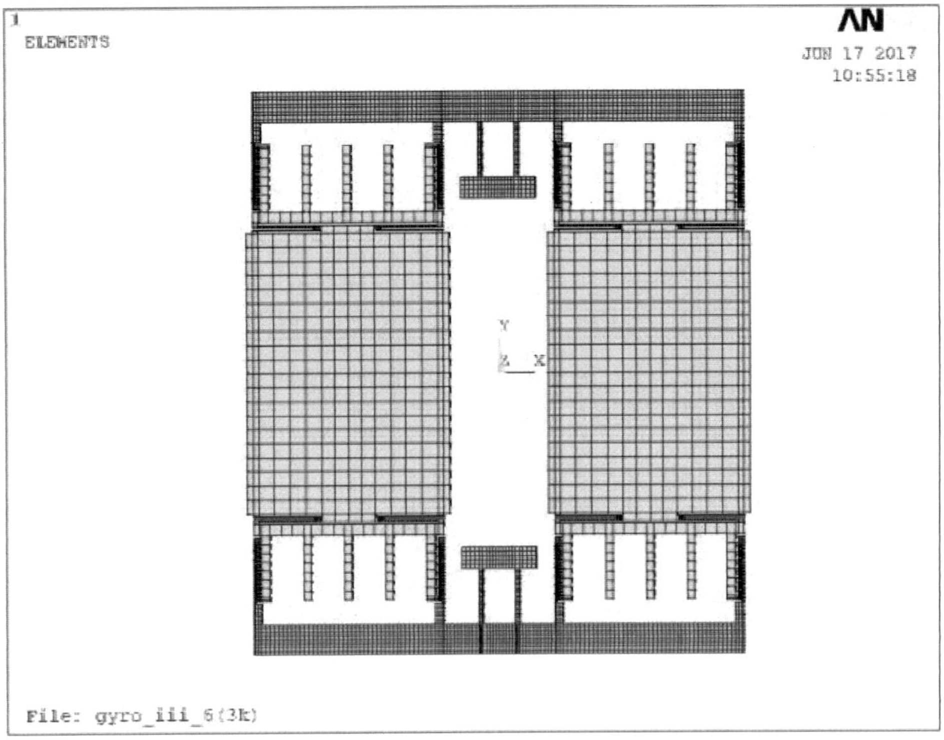

Figure 3. The schematic of a dual mass silicon micro-machined gyroscope.

Table 2. The natural frequency variation.

	theoretic frequency change (Hz)	Simulated frequencychange (Hz)	Error (%)
In-phase drive mode	0.49	0.43	13.9
Anti-phase drive mode	0.55	0.58	5.1
In-phase sense mode	1.54	1.38	11.5
Anti-phase sense mode	1.52	1.45	4.8

Then the harmonic response simulation is done to verify the modal coupling coefficient. Fig. 4 shows the frequency response of the anti-phase drive mode. The frequency response of other modes has the similar results. Fig. 4 shows the in-phase and anti-phase displacements are not independent from each other. Fig. 5 shows the modal coupling coefficient of theory and simulation in each mode. According to Fig. 5, with the stiffness difference of 1%, the modal coupling coefficient is 12% in the driving direction and 31% in the sensing direction. It also shows that the theoretical result and simulation result are in good agreement.

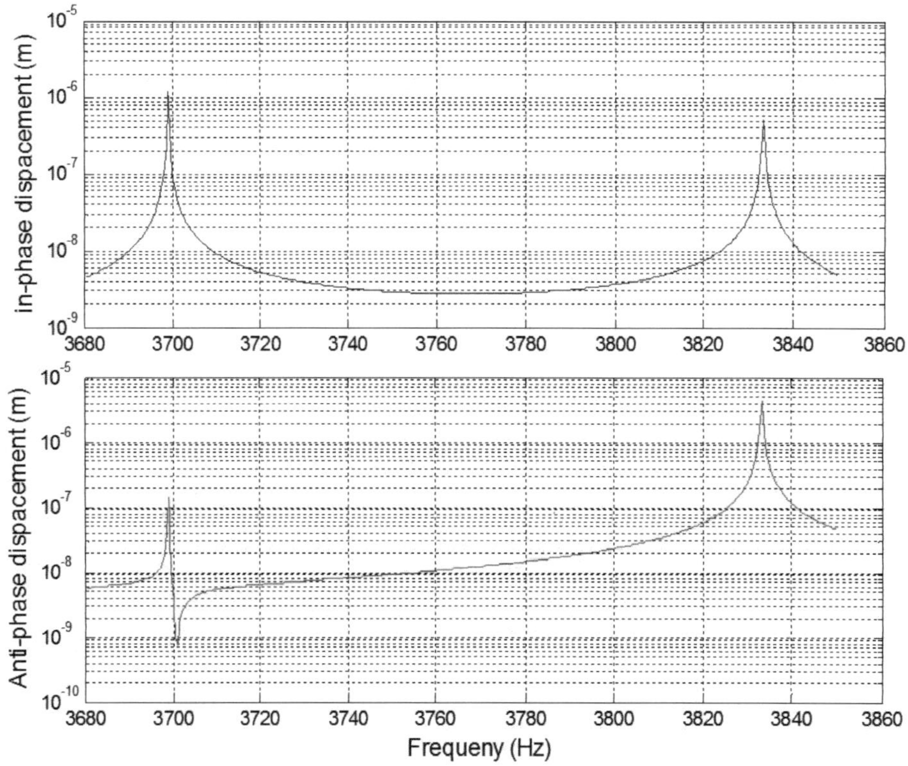

Figure 4. The frequency response of the anti-phase drive mode.

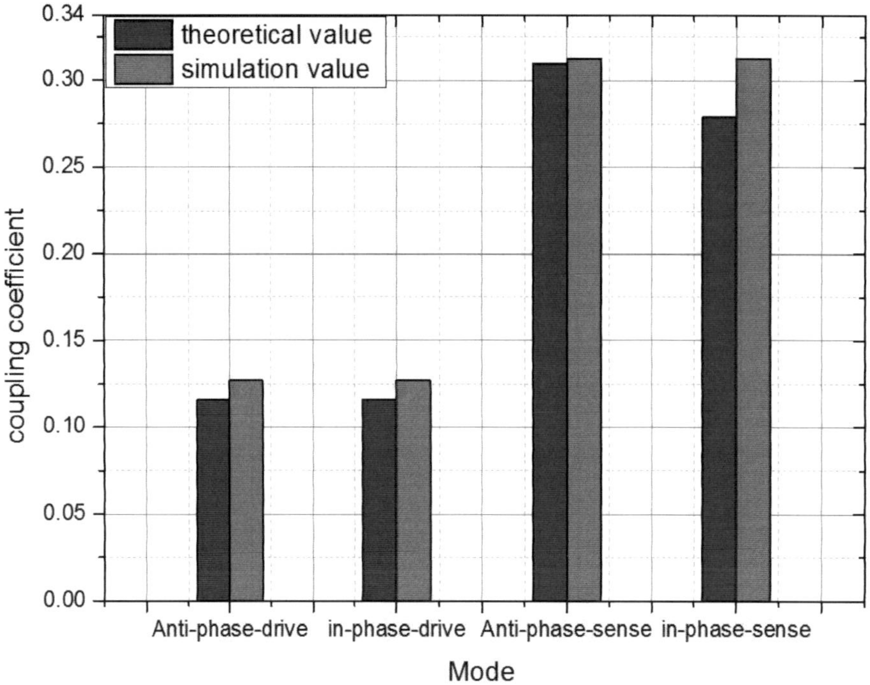

Figure 5. The mode coupling coefficient of theory and simulation in each mode(the theoretical and the simulation value are located in the left and right of each column).

5. Conclusion

Base on the structural characteristics, a coupling dynamic model is proposed and established for the dual mass silicon micro-gyroscope in this paper. The effect of the stiffness difference on the natural frequencies and mode functions are demonstrated. The modal coupling coefficient is calculated based on the mode functions. The natural frequency variations due to the stiffness difference of 1% are 0.5Hz in its driving direction and 1.5Hz in its sensing direction. The modal coupling coefficient is 12% in the drive direction and 31% in the sense direction due to the stiffness difference of 1%. These theoretical results are verified by the FEM simulation. The good agreement between the theory result and simulation result prove the accuracy of our model and its theory.

Acknowledgement

This work is supported by Department of Instrument Science and Engineering at Nanjing University of Science and Technology. The authors also thank the personnel of Department of Instrument Science and Engineering at Nanjing University of Science and Technology for designing, fabricating, and packing the device for this study.

References

[1] M S Weinberg and A Kourepenis Jun. 2006 "Error sources in in-plane silicon tuning-fork MEMS gyroscopes," *J. Microelectromech. Syst.* vol **15** no. 3 pp 479–91
[2] S W Yoon, S W Lee, N C Perkins and K Najafi Oct. 28–31, 2007 "Vibration sensitivity of MEMS tuning fork gyroscopes," in *Proc. IEEE Sensors*, Atlanta GA, USA pp. 115-19
[3] L Huang, Y Ni and Y Yin Sept. 16-18, 2011 "Fabrication error analysis on dual-mass silicon micro-gyroscope," *International Conference on Electrical and Control Engineering* pp 3931-34
[4] A R Schofield, A A Trusov and A M Shkel Oct. 28–31, 2007 "Multi-degree of freedom tuning fork gyroscope demonstrating shock rejection," in *Proc. IEEE Sensors* Atlanta GA USA pp 120–23
[5] B Yang, X Wang, D Hu and L Wu 2017 "Research on the non-ideal dynamics of a dual-mass silicon micro-gyroscope," *Microsystem Technologies,* vol **23** no 1 pp 151-62
[6] B Yang, D Hu, Y Deng and X Wang Feb. 2016 "An improved dual-mass decoupled micro-gyroscope for the non-ideal decoupled error suppression". in *Proc. IEEE Int. Symp. Inertial Sens. Syst* pp 58-61

NOx Pollution Analysis for a Sulfur Recovery Unit Thermal Reactor

Chun-Lang Yeh

Department of Aeronautical Engineering, National Formosa University, Taiwan

E-mail: clyeh@nfu.edu.tw

Abstract. A sulfur recovery unit (SRU) thermal reactor is the most important equipment in a sulfur plant. It is negatively affected by high temperature operations. In this paper, NOx emissions from the SRU thermal reactors are simulated. Both the prototype thermal reactor and its modifications, including changing fuel mass fraction, changing inlet air quantity, changing inlet oxygen mole fraction, and changing burner geometry, are analyzed to investigate their influences on NOx emissions. In respect of the fuel mass fraction, the simulation results show that the highest NO emission occurs at a zone 1 fuel mass fraction of 0.375, around which the reactor maximum temperature and the zone 1 average temperature reach maximum values. Concerning the inlet air quantity, the highest NO emission occurs when the inlet air quantity is 2.4 times the designed inlet air quantity. This is very close to the inlet air quantity at which the maximum average temperature occurs. Regarding the inlet oxygen mole fraction, the NO emission increases as the inlet oxygen mole fraction increases. With regard to the burner geometry, the NO emission increases as the clearance of the burner acid gas tip increases. In addition, the NO emission increases as the swirling strength increases.

1. Introduction

A SRU thermal reactor is the most important equipment in a sulfur plant. It converts the NH_3, H_2S and hydrocarbons in the reactants into sulfur. Most of the sulfur elements are recovered from the SRU thermal reactor. The first section of a SRU that uses the Claus process is composed of a burner, a thermal reactor and a waste heat exchanger. The configuration and dimensions of the first section of a SRU for a typical petroleum refinery are shown in figure 1.

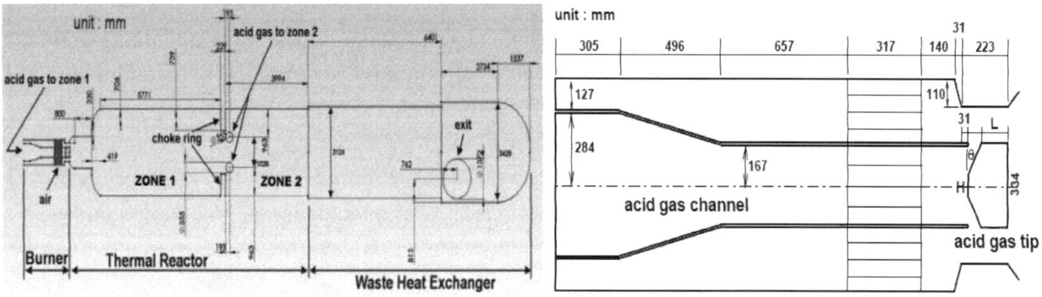

(a) the overall view (b) enlarged view for the burner section

Figure 1. The configuration and dimensions of the first section of the SRU for a typical petroleum refinery.

SRU thermal reactors are negatively affected by high temperature operations because high temperature can damage the refractory and the heat exchanger tubes. Our experience of operating a practical SRU thermal reactor in Taiwan shows that the refractories at the zone 1 corner and the choke ring are the parts of the thermal reactor that experience the greatest deterioration. The zone 1 corner has a suddenly expanded geometry and a recirculation zone forms behind it. The temperature at the zone 1 corner can exceed the maximum service temperature of the refractory and cause collapse or deformation. The choke ring is subjected to a bending moment from the rapid combustion gas stream and can collapse or deform.

Nitrogen oxide (NOx) is one of the major sources of air pollution. It is generated mainly from the high temperature combustion processes and may result in acid rain and cause damage to the atmospheric ozone layer. In addition, NOx has been recognized as one of the major causes of excessive peroxide concentration in the atmosphere. Thus, there is a common consensus that NOx control has become a worldwide problem. Normann et al. [1] reviewed available techniques for controlling both the emission of NOx to the atmosphere and the content of NOx in the captured carbon dioxide. They indicated that for a first generation of oxy-fuel power plants, conventional primary NOx control should be sufficient to meet today's emission regulations, if based on emission per unit of fuel supplied. However, there are several opportunities for new methods of NOx control in oxy-fuel plants, depending on future emission and storage legislation for carbon capture schemes. Javed et al. [2] presented a review of NOx removal techniques with particular reference to SNCR technology. A review of various features related to selective non-catalytic gas phase injection of ammonia and ammonium salts (as reducing agent) was presented. Computational fluid dynamics (CFD) modeling was also applied to SNCR. In addition, a two-stage NOx removal strategy to control un-reacted ammonia slip and to improve overall efficiency was discussed and a summary highlighted various areas needing further research was given. Other reviews for NOx emissions and controls can be found in the author's previous study [3].

In this paper, NOx emissions from the SRU thermal reactors are simulated. Practical operating conditions from a petrochemical corporation in Taiwan are used as the designed conditions for the discussion. Both the prototype SRU thermal reactor and its modifications, including changing fuel mass fraction, changing inlet air quantity, changing inlet oxygen mole fraction, and changing burner geometry, are analyzed to investigate their influences on NOx emissions.

2. Numerical methods and physical models

In this study, the FLUENT commercial code is used to simulate the reacting and fluid flow in a SRU thermal reactor. The SIMPLE algorithm by Patankar [4] is used to solve the governing equations. The discretizations of convection terms and diffusion terms are respectively performed using the power-law scheme and the central difference scheme. In terms of physical models, considering the accuracy and stability of the models and the evaluations of other researchers, the standard k-ε Model [5], the P-1 radiation model [6] and the non-premixed combustion model with β–type probability density function [7] are respectively used for the turbulence, radiation and combustion simulations. The standard wall functions [8] are used to resolve the flow quantities (velocity, the temperature, and the turbulence quantities) at the near-wall regions. The detailed governing equations and convergence criterion were described in a previous study by the author [9].

In this study, the numerical model of a SRU thermal reactor is constructed using an unstructured grid. Figure 2 shows the numerical model of the prototype SRU thermal reactor. In figure 2, the heat exchanger section consists of 19 cooling tubes, which have a diameter of 0.5m, as shown schematically in figure 3. The heat absorption rate for each heat exchanger tube is 40,000 W/m^2 and the other walls are adiabatic. No slip condition is applied on any of the solid walls. The exit of the heat exchanger section is connected to other equipment at 300 K and 1 atm by a pipe that is 1.372m in diameter and 11.5m in length.

Figure 2. The numerical model for the SRU thermal reactor investigated.

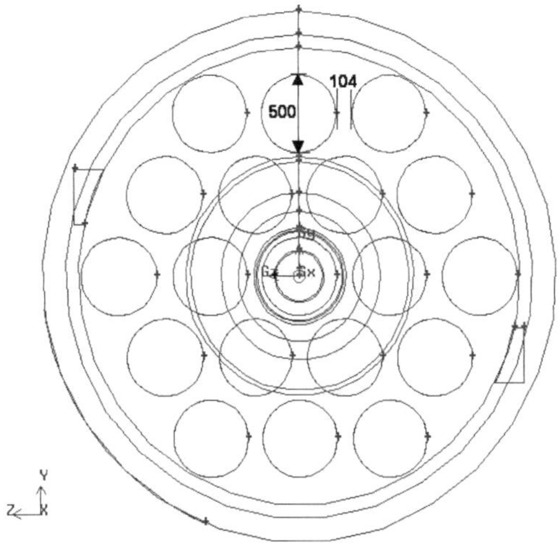

Figure 3. An illustration of the arrangement of heat exchanger tubes.

In this study, two types of oxygen supplies are investigated: an oxygen-normal supply and an oxygen-rich supply. An oxygen-rich supply increases sulfur recovery. The designed conditions (including the species compositions, the temperature, the pressure and the velocity) at the acid gas inlet holes of zone 1 and zone 2 and at the air inlet hole are listed in the author's previous studies[10,11]. These conditions are practical operating conditions that are used by a petrochemical corporation in Taiwan. The turbulence kinetic energy is 10% of the inlet mean flow kinetic energy and the turbulence dissipation rate is computed using Eq.(1).

$$\varepsilon = C_\mu^{3/4} \frac{k^{3/2}}{l} \tag{1}$$

where l=0.07L and L is the hydraulic diameter.

3. Results and discussion

3.1. Prototype SRU thermal reactor

The exit NO concentration, maximum temperature and average temperature for the prototype SRU thermal reactor are listed in table 1. This result will be used as the basis for comparison in the subsequent discussions.

Table 1. Exit NO concentration, maximum temperature and average temperature for the prototype SRU thermal reactor.

NO_{exit}	1.18	14.7
T_{max}	1931	2084
T_{avg}	1548	1598

3.2. Effect of fuel mass fraction

To investigate the effect of fuel mass fraction on NO emission, the zone 1 fuel mass fraction is defined as the mass ratio of the zone 1 acid gas to the total acid gas (acid gas to zone 1 plus acid gas to zone 2). The designed zone 1 fuel mass fractions are 0.875 for an oxygen-normal supply and 0.895 for an oxygen-rich supply.

Figure 4 shows the exit NO concentrations for SRU thermal reactors with different zone 1 fuel mass fractions. The air is kept at the designed condition. Thermal NOx arises from the nitrogen in air when the temperature exceeds 1600°C. From the author's previous study [10], it was found that the maximum temperature and the average temperature reach maximum values at a zone 1 fuel mass fraction of about 0.375, as shown in figure 5, which implies that the stoichiometric fuel mass fraction is around 0.375. The temperature decreases when the zone 1 fuel mass fraction deviates from the stoichiometric fuel mass fraction. It can be observed from figure 4 that the highest NO emission occurs at a zone 1 fuel mass fraction of 0.375, around which the reactor maximum temperature and the zone 1 average temperature reach maximum values.

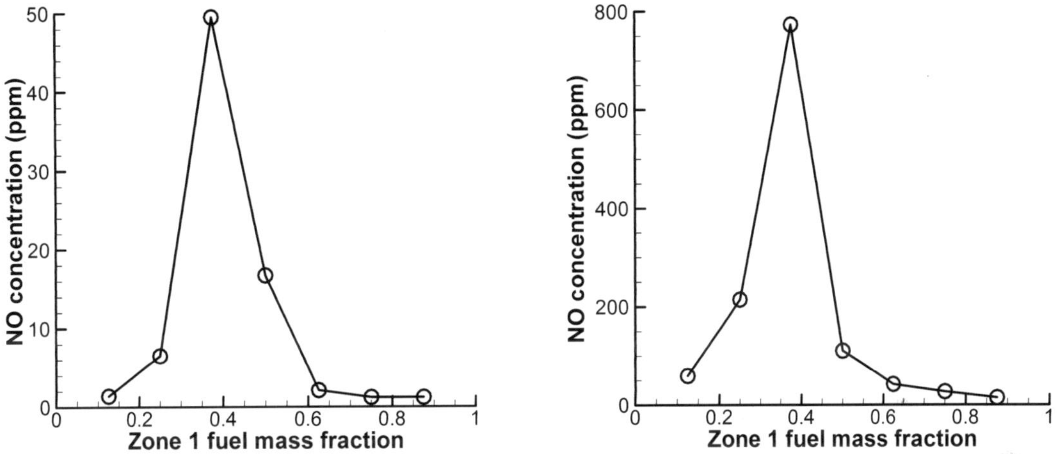

(a) using oxygen-normal supply (b) using oxygen-rich supply

Figure 4. Exit NO concentrations for SRU thermal reactors with different zone 1 fuel mass fractions.

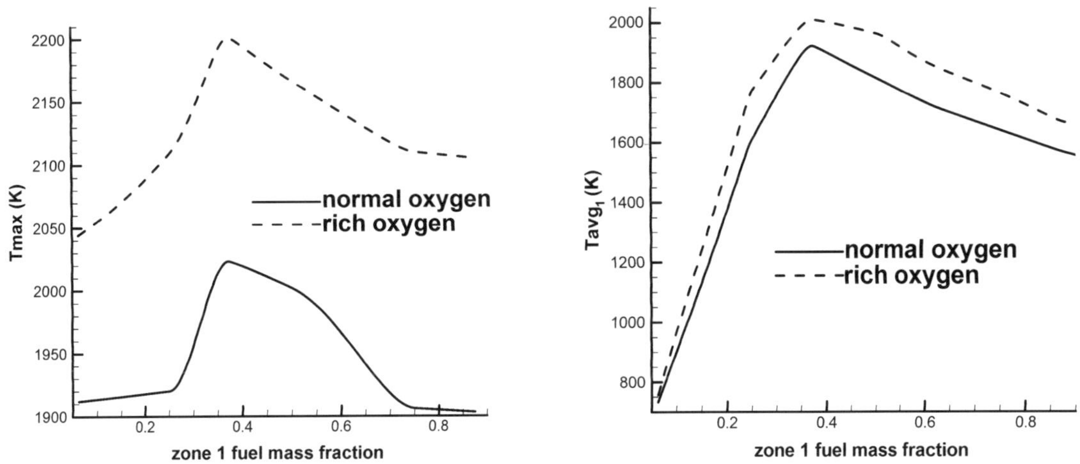

(a) maximum temperature (b) the average temperature in zone 1

Figure 5. A comparison of the maximum and zone 1 average temperatures for SRU thermal reactors with different zone 1 fuel mass fractions.

3.3. Effect of inlet air quantity

To investigate the effect of inlet air quantity on NO emission, 25 inlet air quantities ranging from 0.1-2.5 times the designed inlet air quantity are investigated. The fuel (acid gas to zone 1 and zone 2) is maintained at the designed conditions.

Figure 6 shows the exit NO concentrations for SRU thermal reactors with different inlet air quantities. From the author's previous study [11], it was found that the temperature increases as the inlet air quantity increases, until a maximum temperature is reached at about 2.3 times the designed inlet air quantity, as shown in figure 7. This demonstrates that the designed conditions are fuel-rich (or air-lean) conditions. From figure 6, it is observed that the highest NO emission occurs when the inlet air quantity is 2.4 times the designed inlet air quantity. This is very close to the inlet air quantity at which the maximum average temperature occurs.

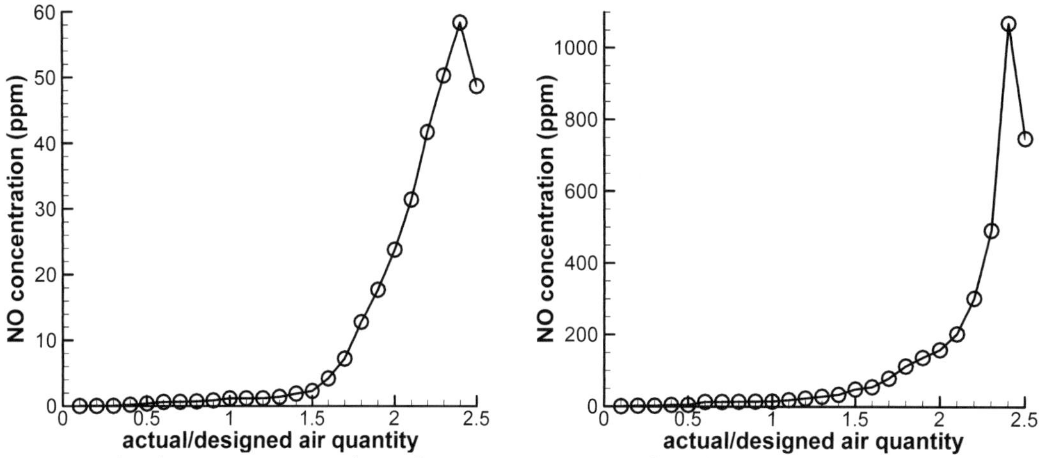

(a) using oxygen-normal supply (b) using oxygen-rich supply

Figure 6. Exit NO concentrations for SRU thermal reactors with different inlet air quantities.

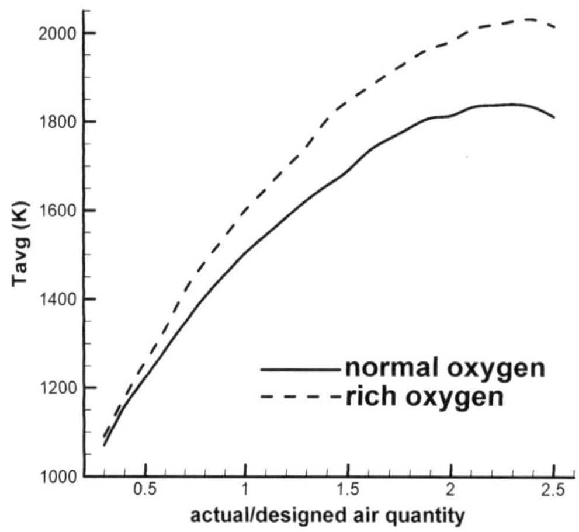

Figure 7. A comparison of the average temperature for a SRU thermal reactor with different inlet air quantities.

3.4. Effect of inlet oxygen mole fraction

Figure 8 shows the exit NO concentrations for SRU thermal reactors with different inlet O_2 mole fractions. The inlet O_2 mole fractions range from 0.16-0.25. For inlet O_2 mole fractions less than or equal to 0.2 (oxygen-lean), the inlet air quantity is equal to the oxygen-normal case. For inlet O_2 mole fractions greater than 0.2 (oxygen-rich), the inlet air quantity is equal to the oxygen-rich case. The fuel (acid gas to zone 1 and zone 2) conditions are kept at the designed conditions. From the author's previous study [11], it was found that the temperature increases monotonically as the inlet O_2 mole fraction increases because the chemical reaction is more complete, as shown in figure 9. The average temperature for an inlet O_2 mole fraction of 0.21 is lower than that for an inlet O_2 mole fraction of 0.2 because the inlet air quantity for the oxygen-rich case is less than that for the oxygen-normal case.

From figure 8, it is observed that the NO emission increases as the inlet O_2 mole fraction increases. Because the thermal NOx arises from the nitrogen in air when the temperature exceeds 1600°C, the NO emission increases dramatically as the inlet O_2 mole fractions are greater than 0.23 at which the average temperatures in zone 1 are above 1600°C.

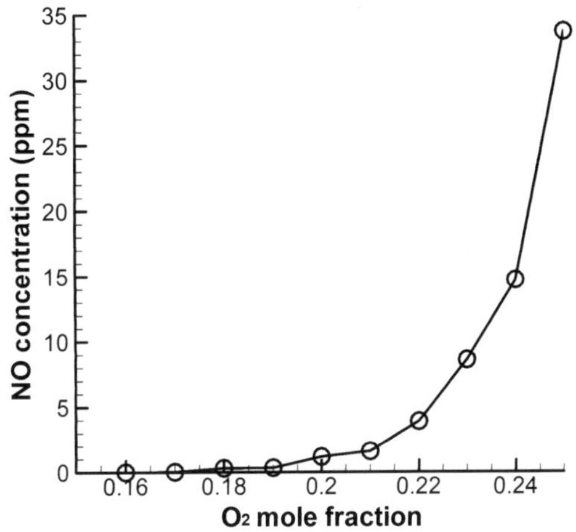

Figure 8. Exit NO concentrations for SRU thermal reactors with different inlet O_2 mole fractions.

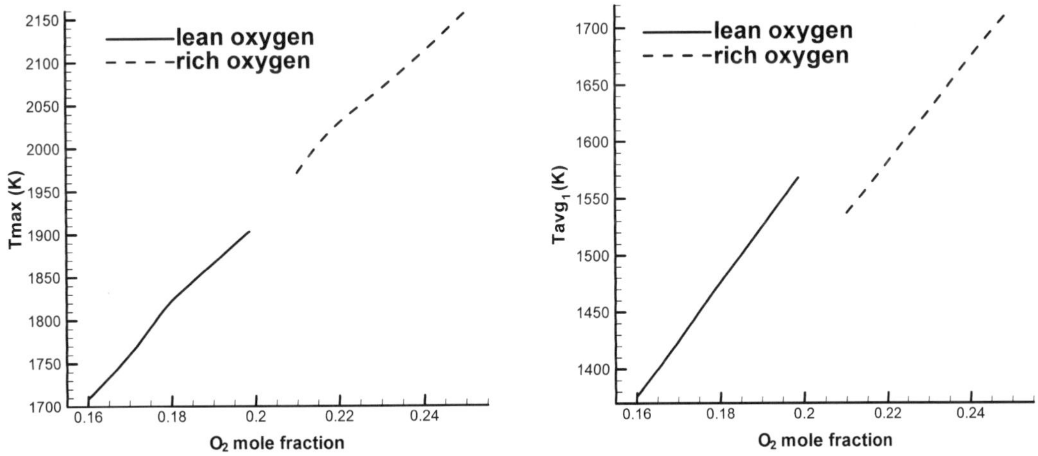

(a) maximum temperature (b) the average temperature in zone 1

Figure 9. A comparison of the temperature for a SRU thermal reactor with different inlet O_2 mole fractions.

3.5. *Effect of burner parameters*

To investigate the effect of burner parameters on NO emission, the clearance of the acid gas tip and the inlet air swirler angle are investigated. The fuel (acid gas to zone 1 and zone 2) and air are kept at the designed conditions.

Figure 1(b) shows the configuration and dimension of the burner section, from which it is seen that length of the acid gas tip (L) controls the horizontal clearance between the acid gas tip and the acid gas channel. The horizontal clearance decreases with L. Figure 10 shows the exit NO concentrations for

SRU thermal reactors with different horizontal clearances of the acid gas tip. The vertical clearance is kept constant by setting the height of the acid gas tip (H) at 80mm. Six lengths of the acid gas tip (L) are used and compared, including 25.4mm, 50.8mm, 76.2mm, 101.6mm, 127mm and 152.4mm, among which 127mm is the default L of a petrochemical corporation in Taiwan. From the author's previous study [12], it was found that a larger horizontal clearance (smaller L) tends to yield a higher temperature, as shown in figure 11. Slower ejected acid gas produces a better mixing with the swirling air. In contrast, faster ejected acid gas may retard the mixing with the swirling air. The smallest horizontal clearance (L=152.4mm) leads to the lowest temperature.

From figure 10, it is observed that the NO emission increases as the horizontal clearance increases (L decreases) because a larger horizontal clearance (smaller L) yields a higher temperature.

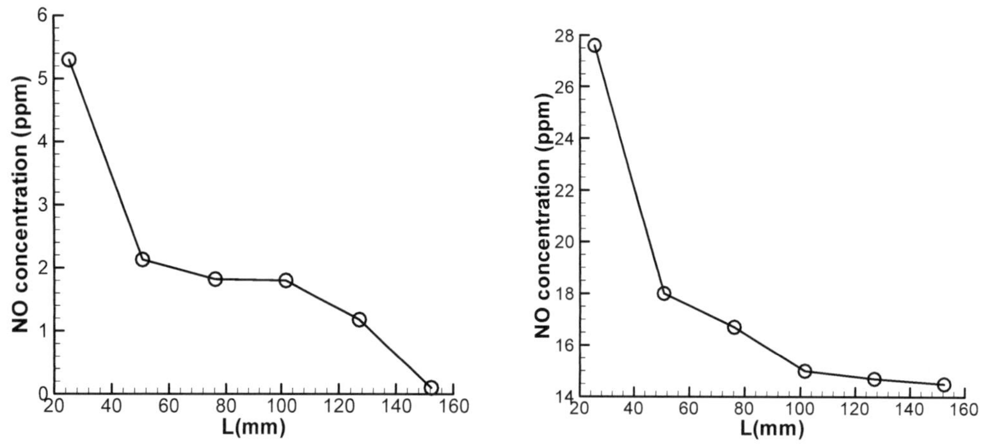

(a) oxygen-normal supply (b) oxygen-rich supply

Figure 10. Exit NO concentrations for SRU thermal reactors with different horizontal clearances of the acid gas tip.

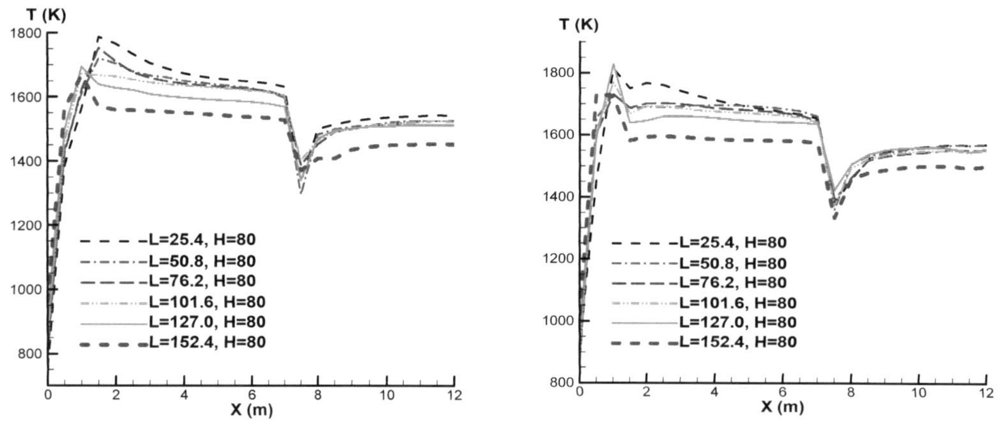

(a) oxygen-normal supply (b) oxygen-rich supply

Figure 11. A comparison of the cross-sectional average temperatures for SRU thermal reactors with different horizontal clearances of the acid gas tip.

Figure 12 shows the exit NO concentrations for SRU thermal reactors with different vertical clearances of the acid gas tip. The horizontal clearance is kept constant by setting the length of the acid gas tip (L) to 127mm. In figure 1(b), the height of the acid gas tip (H) controls the vertical clearance of the acid gas tip. The vertical clearance decreases with H. Four heights of the acid gas tip (H) are used and compared, including 40mm, 80mm, 160mm and 240mm, among which 80mm is the default H of a petrochemical corporation in Taiwan. From the author's previous study [12], it was found that a larger vertical clearance (smaller H) yields a higher temperature, as shown in figure 13. This result is similar to that for a larger horizontal clearance. The smallest vertical clearance (H=240mm) leads to the lowest temperature.

From figure 12, it is observed that the NO emission increases as the vertical clearance increases (H decreases) because a larger vertical clearance (smaller H) yields a higher temperature. This result is similar to that for a larger horizontal clearance.

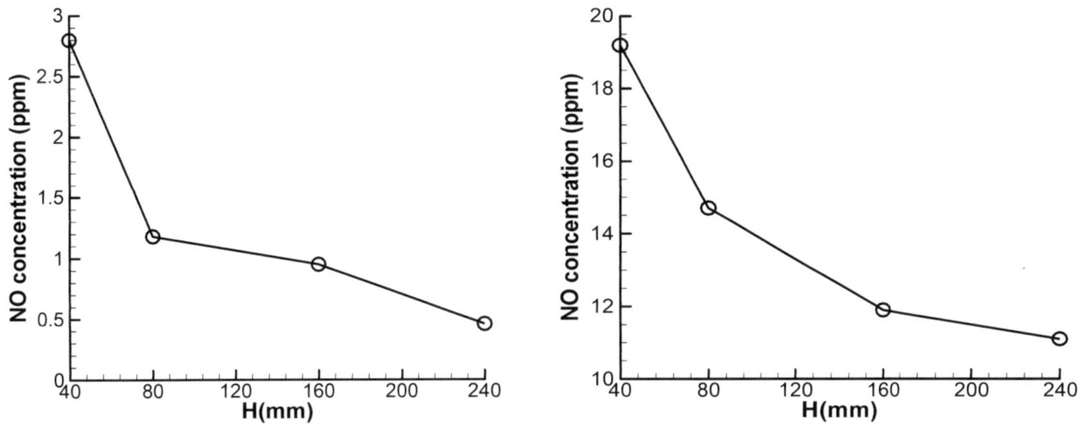

(a) oxygen-normal supply (b) oxygen-rich supply

Figure 12. Exit NO concentrations for SRU thermal reactors with different vertical clearances of the acid gas tip.

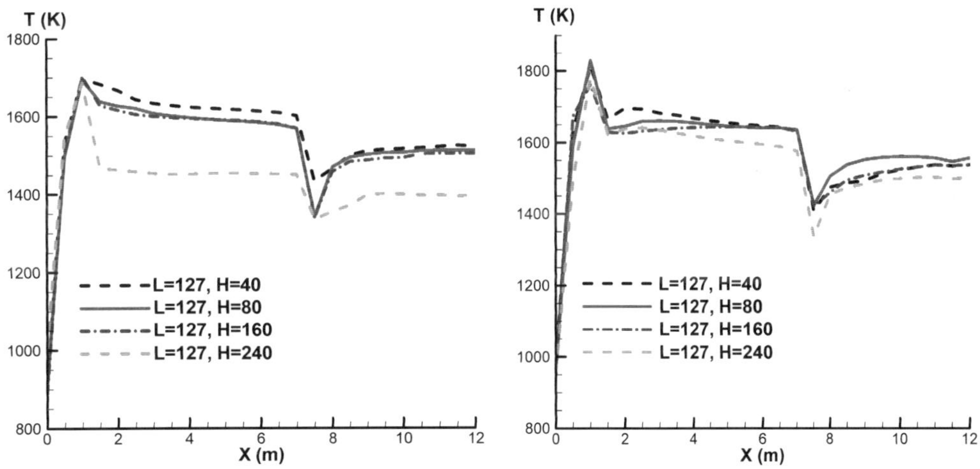

(a) oxygen-normal supply (b) oxygen-rich supply

Figure 13. A comparison of the cross-sectional average temperatures for a SRU thermal reactor with different vertical clearances of the acid gas tip.

Figure 14 shows the exit NO concentrations for SRU thermal reactors with different air swirler angles. The inlet air swirler angle, θ_{sw}, controls the strength of the swirling motion. The swirling strength can be expressed in terms of the radial (v_r) and tangential (v_θ) components of the inlet air velocity by the following equation.

$$v_r / v_\theta = \tan \theta_{sw} \quad (2)$$

A smaller θ_{sw} corresponds to a stronger swirling motion. The length (L) and height (H) of the acid gas tip are kept at 127mm and 80mm, respectively. Four inlet air swirler angles (θ_{sw}) are used and compared, including 10°, 20°, 30° and 60°. From the author's previous study [12], it was found that a smaller θ_{sw} yields a higher zone I temperature, as shown in figure 15. This is because a stronger swirling motion leads to a better mixing, a more complete chemical reaction and hence a higher temperature. The smallest θ_{sw} (10°) produces the highest zone I temperature.

From figure 14, it is observed that the NO emission increases as the swirling strength increases (swirler angle, θ_{sw}, decreases) because a stronger swirling motion (smaller θ_{sw}) yields a higher temperature.

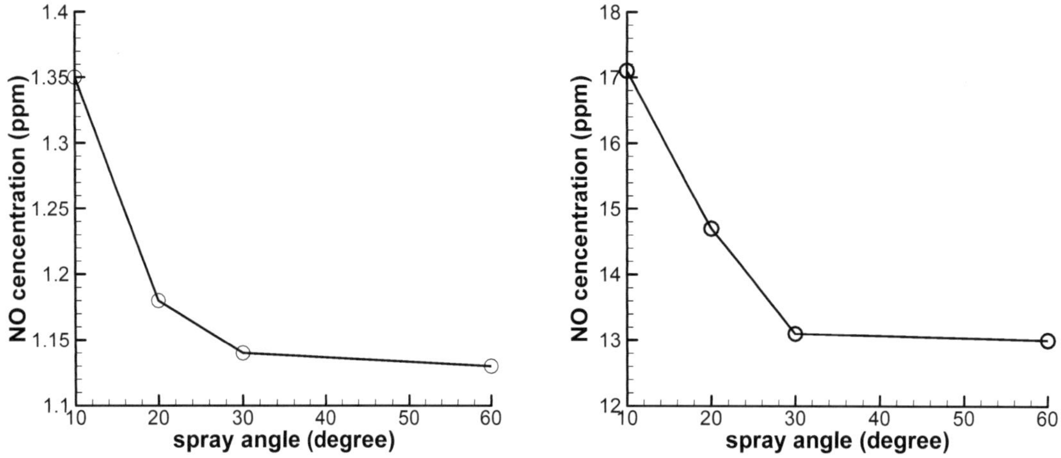

(a) oxygen-normal supply (b) oxygen-rich supply

Figure 14. Exit NO concentrations for SRU thermal reactors with different air swirler angles.

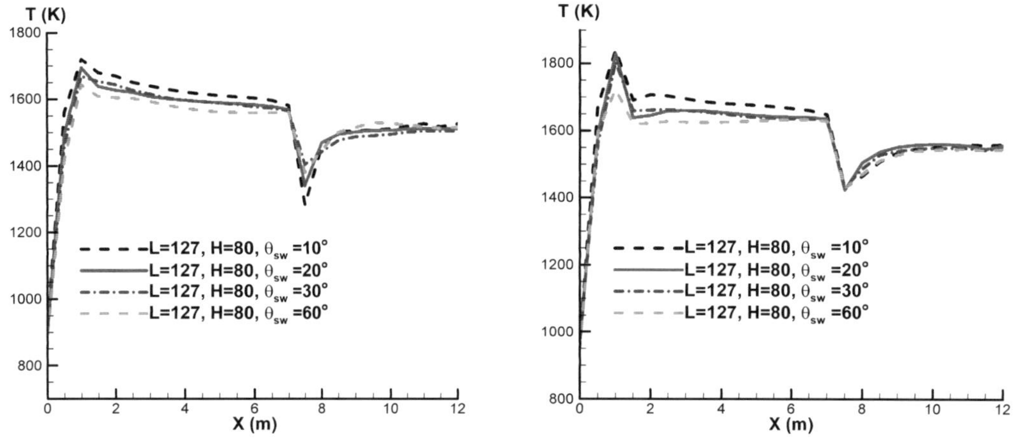

(a) oxygen-normal supply (b) oxygen-rich supply

Figure 15. A comparison of the cross-sectional average temperatures for SRU thermal reactors with different inlet air swirler angles.

4. Conclusion

In this paper, NOx emissions from the SRU thermal reactors are simulated. Practical operating conditions from a petrochemical corporation in Taiwan are used as the designed conditions for the discussion. Both the prototype SRU thermal reactor and its modifications, including changing fuel mass fraction, changing inlet air quantity, changing inlet oxygen mole fraction, and changing burner geometry, are analyzed to investigate their influences on NOx emissions. From the simulation results, the following conclusions may be drawn:

(1) In respect of the fuel mass fraction, the highest NO emission occurs at a zone 1 fuel mass fraction of 0.375, around which the reactor maximum temperature and the zone 1 average temperature

reach maximum values.

(2) Concerning the inlet air quantity, the highest NO emission occurs when the inlet air quantity is 2.4 times the designed inlet air quantity. This is very close to the inlet air quantity at which the maximum average temperature occurs.

(3) Regarding the inlet oxygen mole fraction, the NO emission increases as the inlet oxygen mole fraction increases.

(4) With regard to the burner geometry, the NO emission increases as the clearance of the burner acid gas tip increases. In addition, the NO emission increases as the swirling strength increases.

References

[1] Normann F, Anderson K., Leckner B and Johnson F 2009 "Emission Control of Nitrogen Oxides in the Oxy-Fuel Process", *Progress in Energy and Combustion Science* Vol.**35** Issue 5 pp 385-97

[2] Javed M T, Irfana N and Gibbs B M 2007 "Control of Combustion-generated Nitrogen Oxides by Selective Non-catalytic Reduction", *Journal of Environmental Management* Vol.**83** Issue 3 pp 251-89

[3] Yeh C L 2013 "Numerical Study of the Influence of Refractory Thickening on the Reacting Flow and DeNOx Effect in a CO Boiler", *Applied Mechanics and Materials* Vol.**284-287** pp 861-66

[4] Patankar S V 1980 *Numerical Heat Transfer and Fluid Flows*, McGraw-Hill New York USA

[5] Launder B E, Spalding D B 1972 *Lectures in Mathematical Models of Turbulence*, Academic Press, London, England

[6] Siegel R and Howell, J R 1992 *Thermal Radiation Heat Transfer*, Hemisphere Publishing Corporation Washington DC USA

[7] Sivathanu Y R and Faeth G M 1990 "Generalized State Relationships for Scalar Properties in Non-premixed Hydrocarbon/air Flames", *Combustion and Flame* Vol.**82** pp 211-30

[8] Launder B E and Spalding D B 1974 "The Numerical Computation of Turbulent Flows", *Computer Methods in Applied Mechanics and Engineering* Vol.**3** No.2 pp 269-89

[9] Yeh C L 2013 "Numerical Analysis of the Combustion and Fluid Flow in a Carbon Monoxide Boiler", *International Journal of Heat and Mass Transfer* Vol.**59** pp 172-90

[10] Yeh C L 2016 "The Effect of Fuel Mass Fraction on the Combustion and Fluid Flow in a Sulfur Recovery Unit Thermal Reactor", *Applied Sciences* Vol.**6** Issue 11 pp 331 doi:10.3390/app6110331

[11] Yeh C L 2017 "The Effects of Inlet Air Quantity and Inlet Oxygen Mole Fraction on the Combustion and Fluid Flow in a Sulfur Recovery Unit Thermal Reactor", *Transactions of the Canadian Society for Mechanical Engineering* Vol.**41** No.2 pp 293-300

[12] Yeh C L 2017 "Numerical Study of the Burner Parameters on the Thermal Field in a Sulfur Recovery Unit Thermal Reactor", *MATEC Web of Conferences*